BIG DATA

BIG DATA

How the Information Revolution is Transforming Our Lives

BRIAN CLEGG

ICON

Published in the UK in 2017
by Icon Books Ltd, Omnibus Business Centre,
39–41 North Road, London N7 9DP
email: info@iconbooks.com
www.iconbooks.com

Sold in the UK, Europe and Asia
by Faber & Faber Ltd, Bloomsbury House,
74–77 Great Russell Street,
London WC1B 3DA or their agents

Distributed in the UK, Europe and Asia
by Grantham Book Services,
Trent Road, Grantham NG31 7XQ

Distributed in the USA
by Publishers Group West,
1700 Fourth Street, Berkeley, CA 94710

Distributed in Australia and New Zealand
by Allen & Unwin Pty Ltd,
PO Box 8500, 83 Alexander Street,
Crows Nest, NSW 2065

Distributed in South Africa
by Jonathan Ball, Office B4, The District,
41 Sir Lowry Road, Woodstock 7925

Distributed in India by Penguin Books India,
7th Floor, Infinity Tower – C, DLF Cyber City,
Gurgaon 122002, Haryana

Distributed in Canada by Publishers Group Canada,
76 Stafford Street, Unit 300
Toronto, Ontario M6J 2S1

ISBN: 978-178578-234-3

Typeset in Iowan by Marie Doherty

Printed and bound in the UK
by Clays Ltd, St Ives plc

ABOUT THE AUTHOR

Brian Clegg's most recent books are *The Reality Frame* (Icon, 2017), *What Colour is the Sun* (Icon, 2016) and *Ten Billion Tomorrows* (St Martin's Press, 2016). His *Dice World* and *A Brief History of Infinity* were both longlisted for the Royal Society Prize for Science Books. Brian has written for numerous publications including *The Wall Street Journal*, *Nature*, *BBC Focus*, *Physics World*, *The Times*, *The Observer*, *Good Housekeeping* and *Playboy*. Brian is editor of popularscience.co.uk and blogs at brianclegg.blogspot.com.

www.brianclegg.net

For Gillian, Chelsea and Rebecca

ACKNOWLEDGEMENTS

I've had a long relationship with data and information. When I was at school we didn't have any computers, but patient teachers helped us to punch cards by hand, which were sent off by post to London and we'd get a print-out about a week later. This taught me the importance of accuracy in coding – so thanks to Oliver Ridge, Neil Sheldon and the Manchester Grammar School. I also owe a lot to my colleagues at British Airways, who took some nascent skills and turned me into a data professional; particular mention is needed for Sue Aggleton, John Carney and Keith Rapley. And, as always, thanks to the brilliant team at Icon Books who were involved in producing this series, notably Duncan Heath, Simon Flynn, Robert Sharman and Andrew Furlow.

CONTENTS

WE KNOW WHAT YOU'RE THINKING

1

The big deal about big data

It's hard to avoid 'big data'. The words are thrown at us in news reports and from documentaries all the time. But we've lived in an information age for decades. What has changed?

Take a look at a success story of the big data age: Netflix. Once a DVD rental service, the company has transformed itself as a result of big data – and the change is far more than simply moving from DVDs to the internet. Providing an on-demand video service inevitably involves handling large amounts of data. But so did renting DVDs. All a DVD does is store gigabytes of data on an optical disc. In either case we're dealing with data processing on a large scale. But big data means far more than this. It's about making use of the whole spectrum of data that is available to transform a service or organisation.

Netflix demonstrates how an on-demand video company can put big data at its heart. Services like Netflix involve more two-way communication than a conventional broadcast. The company knows who is watching what, when and where. Its systems can cross-index measures of a viewer's interests, along with their feedback. We as viewers see the outcome of this analysis in the recommendations Netflix makes, and sometimes they seem odd, because the system is attempting to predict the likes and dislikes of a single individual. But from the Netflix viewpoint, there is a much greater and more effective benefit in matching preferences across large populations: it can transform the process by which new series are commissioned.

Take, for instance, the first Netflix commission to break through as a major series: *House of Cards*. Had this been a project for a conventional network, the broadcaster would have produced a pilot, tried it out on various audiences, perhaps risked funding a short season (which could be cancelled part way through) and only then committed to the series wholeheartedly. Netflix short-circuited this process thanks to big data.

The producers behind the series, Mordecai Wiczyk and Asif Satchu, had toured the US networks in 2011, trying to get funding to produce a pilot. However, there hadn't been a successful political drama since *The West Wing* finished in 2006 and the people controlling the money felt that *House of Cards* was too high risk. However, Netflix knew from their mass of customer data that they had a large customer base who appreciated the humour and darkness of the original BBC drama the show was based on, which was already in the Netflix library. Equally, Netflix had a lot of customers who liked the work of director David Fincher and actor

Kevin Spacey, who became central to the making of the series.

Rather than commission a pilot, with strong evidence that they had a ready audience, Netflix put $100 million up front for the first two series, totalling 26 episodes. This meant that the makers of *House of Cards* could confidently paint on a much larger canvas and give the series far more depth than it might otherwise have had. And the outcome was a huge success. Not every Netflix drama can be as successful as *House of Cards*. But many have paid off, and even when the takeup is slower, as with the 2016 Netflix drama *The Crown*, given a similar high-cost two-season start, shows have far longer to succeed than when conventionally broadcast. The model has already delivered several major triumphs, with decisions driven by big data rather than the gut feel of industry executives, infamous for getting it wrong far more frequently than they get it right.

The ability to understand the potential audience for a new series was not the only way that big data helped make *House of Cards* a success. Clever use of data meant, for instance, that different trailers for the series could be made available to different segments of the Netflix audience. And crucially, rather than release the series episode by episode, a week at a time as a conventional network would, Netflix made the whole season available at once. With no advertising to require an audience to be spread across time, Netflix could put viewing control in the hands of the audience. This has since become the most common release strategy for streaming series, and it's a model that is only possible because of the big data approach.

Big data is not all about business, though. Among other things, it has the potential to transform policing by predicting likely crime locations; to animate a still photograph; to provide the first ever vehicle for genuine democracy; to predict the next *New York Times* bestseller; to give us an understanding of the fundamental structure of nature; and to revolutionise medicine.

Less attractively, it means that corporations and governments have the potential to know far more about you, whether to sell to you or to attempt to control you. Don't doubt it – big data is here to stay, making it essential to understand both the benefits and the risks.

The key

Just as happened with Netflix's analysis of the potential *House of Cards* audience, the power of big data derives from collecting vast quantities of information and analysing it in ways that humans could never achieve without computers in an attempt to perform the apparently impossible.

Data has been with us a long time. We are going to reach back 6,000 years to the beginnings of agricultural societies to see the concept of data being introduced. Over time, through accounting and the written word, data became the backbone of civilisation. We will see how data evolved in the seventeenth and eighteenth centuries to be a tool to attempt to open a window on the future. But the attempt was always restricted by the narrow scope of the data available and by the limitations of our ability to analyse it. Now, for the first time, big data is opening up a new world. Sometimes it's in a flashy way with computers like Amazon's Echo that we interact with using only speech. Sometimes it's under the surface, as happened with supermarket loyalty cards. What's clear is that the applications of big data are multiplying rapidly and possess huge potential to impact us for better or worse.

How can there be so much latent power in something so basic as data? To answer that we need to get a better feel for what big data really is and how it can be used. Let's start with that 'd' word.

SIZE MATTERS

2

Data is ...

According to the dictionary, 'data' derives from the plural of the Latin 'datum', meaning 'the thing that's given'. Most scientists pretend that we speak Latin, and tell us that 'data' should be a plural, saying 'the data are convincing' rather than 'the data is convincing.' However, the usually conservative *Oxford English Dictionary* admits that using data as a singular mass noun – referring to a collection – is now 'generally considered standard'. It certainly sounds less stilted, so we will treat data as singular.

'The thing that's given' itself seems rather cryptic. Most commonly it refers to numbers and measurements, though it could be anything that can be recorded and made use of later. The words in this book, for instance, are data.

You can see data as the base of a pyramid of understanding:

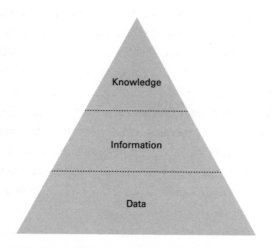

From data we construct information. This puts collections of related data together to tell us something meaningful about the world. If the words in this book are data, the way I've arranged the words into sentences, paragraphs and chapters makes them information. And from information we construct knowledge. Our knowledge is an interpretation of information to make use of it – by reading the book, and processing the information to shape ideas, opinions and future actions, you develop knowledge.

In another example, data might be a collection of numbers. Organising them into a table showing, say, the quantity of fish in a certain sea area, hour by hour, would give you information. And someone using this information to decide when would be the best time to go fishing would possess knowledge.

Climbing the pyramid

Since human civilisation began we have enhanced our technology to handle data and climb this pyramid. This began with clay tablets, used in Mesopotamia at least 4,000 years ago. The tablets allowed data to be practically and useably retained, rather than held in the head or scratched on a cave wall. These were portable data stores. At around the same time, the first data processor was developed in the simple but surprisingly powerful abacus. First using marks or stones in columns, then beads on wires, these devices enabled simple numeric data to be handled. But despite an increasing ability to manipulate data over the centuries, the implications of *big* data only became apparent at the end the nineteenth century as a result of the problem of keeping up with a census.

In the early days of the US census, the increasing quantity of data being stored and processed looked likely to overwhelm the resources available to deal with it. The whole process seemed doomed. There was a ten-year period between censuses – but as population and complexity of data grew, it took longer and longer to tabulate the census data. Soon, a census would not be completely analysed before the next one came round. This problem was solved by mechanisation. Electro-mechanical devices enabled punched cards, each representing a slice of the data, to be automatically manipulated far faster than any human could achieve.

By the late 1940s, with the advent of electronic computers, the equipment reached the second stage of the pyramid. Data processing gave way to information technology. There had been information *storage* since the invention of writing.

A book is an information store that spans space and time. But the new technology enabled that information to be manipulated as never before. The new non-human computers (the term originally referred to mathematicians undertaking calculations on paper) could not only handle data but could turn it into information.

For a long while it seemed as if the final stage of automating the pyramid – turning information into valuable knowledge – would require 'knowledge-based systems'. These computer programs attempted to capture the rules humans used to apply knowledge and interpret data. But good knowledge-based systems proved elusive for three reasons. Firstly, human experts were in no hurry to make themselves redundant and were rarely fully cooperative. Secondly, human experts often didn't know *how* they converted information into knowledge and couldn't have expressed the rules for the IT people even had they wanted to. And finally, the aspects of reality being modelled this way proved far too complex to achieve a useful outcome.

The real world is often chaotic in a mathematical sense. This doesn't mean that what happens is random – quite the opposite. Rather, it means that there are so many interactions between the parts of the world being studied that a very small change in the present situation can make a huge change to a future outcome. Predicting the future to any significant extent becomes effectively impossible.

Now, though, as we undergo another computer revolution through the availability of the internet and mobile computing, big data is providing an alternative, more pragmatic approach to taking on the top level of the data–information–knowledge pyramid. A big data system takes large volumes of data – data that is usually fast flowing and unstructured – and makes use of the latest information technologies to handle and analyse this data in a less rigid, more responsive fashion. Until recently this was impossible. Handling data on this scale wasn't practical, so those who studied a field would rely on samples.

A familiar use of sampling is in opinion polls, where pollsters try to deduce the attitudes of a population from a small subset. That small group is carefully selected (in a good poll) to be representative of the whole population, but there is always assumption and guesswork involved. As recent elections have shown, polls can never provide more than a good guess of the outcome. The 2010 UK general election? The polls got it wrong. The 2015 UK general election? The polls got it wrong. The 2016 Brexit referendum and US presidential election – you guessed it. We'll look at why polls seem to be failing so often a little later (see page 23), but big data gets around the polling problem by taking on everyone – and the technology we now have available means that we can access the data continuously, rather than through the clumsy, slow mechanisms of an old-school big data exercise like a census or general election.

Past, present and future

For lovers of data, each of past, present and future has a particular nuance. Traditionally, data from the past has been the only certainty. The earliest data seems to have been primarily records of past events to support agriculture and trade. It was the bean counters who first understood the value of data. What they worked with then wasn't always very approachable, though, because the very concept of number was in a state of flux.

Look back, for instance, to the mighty city state of Uruk, founded around 6,000 years ago in what is now Iraq. The people of Uruk were soon capturing data about their trades, but they hadn't realised that numbers could be universal. We take this for granted, but it isn't necessarily obvious. So, if you were an Uruk trader and you wanted to count cheese, fresh fish and grain, you would use a totally different number system to someone counting animals, humans or dried fish. Even so, data comes hand in hand with trade, as it does with the establishment of states. The word 'statistics' has the same origin as 'state' – originally it was data about a state. Whether data was captured for trade or taxation or provision of amenities, it was important to know about the past.

In a sense, this dependence on past data was not so much a perfect solution as a pragmatic reflection of the possible. The ideal was to also know about the present. But this was only practical for local transactions until the mechanisms for big data became available towards the end of the twentieth century. Even now, many organisations pretend that the present doesn't exist.

It is interesting to compare the approach of a business driven by big data such as a supermarket with a less data-capable organisation like a book publisher. Someone in the head office of a major supermarket can tell you what is selling across their entire array of shops, minute by minute throughout the day. He or she can instantly communicate demand to suppliers and by the end of the day, the present data is part of the big data source for the next. Publishing (as seen by an author) is very different.

Typically, an author receives a summary of sales for, say, the six months from January to June at the end of September and will be paid for this in October. It's not that on-the-day sales systems don't exist, but nothing is integrated. It doesn't help that publishing operates a data-distorting approach of 'sale or return', whereby books are listed as being 'sold' when they are shipped to a bookstore, but can then be returned for a refund at any time in the future. This is an excellent demonstration of why we struggle to cope with data from the present – the technology might be there, but commercial agreements are rooted in the past, and changing to a big data approach is a significant challenge. And that's just advancing from the past to the present – the future is a whole different ball game.

It wasn't until the seventeenth century that there was a conscious realisation that data collected from the past could have an application to the future. I'm stressing that word 'conscious' because it's something we have always done as humans. We use data from experience to help us prepare for future possibilities. But what was new was to consciously and explicitly use data this way.

It began in seventeenth-century London with a button maker called John Graunt. Out of scientific curiosity, Graunt got his hands on 'bills of mortality' – documents summarising the details of deaths in London between 1604 and 1661.

Graunt was not just interested in studying these numbers, but combined what he could glean from them with as many other data sources as he could – scrappy details, for instance, of births. As a result, he could make an attempt both to see how the population of London was varying (there was no census data) and to see how different factors might influence life expectancy.

It was this combination of data from the past and speculation about the future that helped a worldwide industry begin in London coffee houses, based on the kind of calculations that Graunt had devised. In a way, it was like the gambling that had taken place for millennia. But the difference was that the data was consciously studied and used to devise plans. This new, informed type of gambling became the insurance business. But this was just the start of our insatiable urge to use data to quantify the future.

Crystal balls

There was nothing new, of course, about wanting to foretell what would happen. Who doesn't want to know what's in store for them, who will win a war, or which horse will win the 2.30 at Chepstow? Augurs, astrologers and fortune tellers have done steady business for millennia. Traditionally, though, the ability to peer into the future relied on imaginary mystical powers. What Graunt and the other early statisticians did was offer the hope of a scientific view of the future. Data was to form a glowing chain, linking what had been to what was to come.

This was soon taken far beyond the quantification of life expectancies, useful though that might be for the insurance business. The science of forecasting, the prediction of the data of the future, was essential for everything from meteorology to estimating sales volumes. Forecasting literally means to throw or project something ahead. By collecting data from the past, and as much as possible about the present, the idea of the forecast was to 'throw' numbers into the future – to push aside the veil of time with the help of data.

The quality of such attempts has always been very variable. Moaning about the accuracy of weather forecasts has been a national hobby in the UK since they started in *The Times* in the 1860s, though they are now far better than they were 40 years ago, for reasons we will discover in a moment. We find it very difficult to accept how qualitatively different data from the past and data on the future are. After all, they are sets of numbers and calculations. It all seems very scientific. We have a natural tendency to give each equal weighting, sometimes with hilarious consequences.

Take, for example, a business mainstay, the sales forecast. This is a company's attempt to generate data on future sales based on what has happened before. In every business, on a regular basis, those numbers are inaccurate. And when this happens, companies traditionally hold a post-mortem on 'what went wrong' with their business. This post-mortem process blithely ignores the reality that the forecast, almost by definition, was going to be wrong. What happened is that the forecast did not match the sales, but the post-mortem attempts to establish why the sales did not match the forecast. The reason behind this confusion is a common problem whenever we deal with statistics. We are over-dependent on patterns.

Patterns and self-deception

Patterns are the principal mechanism used to understand the world. Without making deductions from patterns to identify predators and friends, food or hazards, we wouldn't last long. If every time a large object with four wheels came hurtling towards us down a road we had to work out if it was a threat, we wouldn't survive crossing the road. We recognise a car or a lorry, even though we've never seen that specific example in that specific shape and colour before. And we act accordingly. For that matter, science is all about using patterns – without patterns we would need a new theory for every atom, every object, every animal, to explain their behaviour. It just wouldn't work.

This dependence on patterns is fine, but we are so finely tuned to recognise things through pattern that we are constantly being fooled. When the 1976 Viking 1 probe took detailed photographs of the surface of Mars, it sent back an image that our pattern-recognising brains instantly told us was a face, a carving on a vast scale. More recent pictures have shown this was an illusion, caused by shadows when the Sun was at a particular angle. The rocky outcrop bears no resemblance to a face – but it's almost impossible not to see one in the original image. There's even a word for seeing an image of something that isn't there: pareidolia. Similarly,

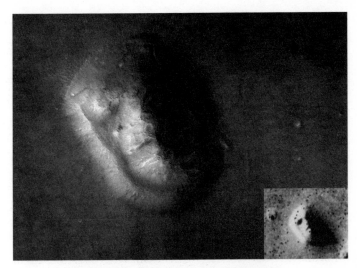

**The 'face on Mars' as photographed in
2001, and *inset* the 1976 image.**

the whole business of forecasting is based on patterns – it is both its strength and its ultimate downfall.

If there are no patterns at all in the historical data we have available, we can't say anything useful about the future. A good example of data without any patterns – specifically designed to be that way – is the balls drawn in a lottery. Currently, the UK Lotto game features 59 balls. If the mechanism of the draw is undertaken properly, there is no pattern to the way these balls are drawn week on week. This means that it is impossible to forecast what will happen in the next draw. But logic isn't enough to stop people trying.

Take a look on the lottery's website and you will find a page giving the statistics on each ball. For example, a table shows how many times each number has been drawn. At the time of writing, the 59-ball draw has been run 116 times. The most frequently drawn balls were 14 (drawn nineteen times) and 41 (drawn seventeen times). Despite there being no connection between them, it's almost impossible to stop a pattern-seeking brain from thinking 'Hmm, that's interesting. Why are the two most frequently drawn numbers reversed versions of each other?'

The least frequent numbers were 6, 48 and 45, each with only five draws each. This is just the nature of randomness. Random things don't occur evenly, but have clusters and gaps. When this is portrayed in a simple, physical fashion it is obvious. Imagine tipping a can of ball bearings on to the floor. We would be very suspicious if they were all evenly spread out on a grid – we expect clusters and gaps. But move away from such an example and it's hard not to feel that there must be a cause for such a big gap between nineteen draws of ball 14 and just five draws of ball 6.

Once such pattern sickness has set in, we find it hard to resist its powerful symptoms. The reason the lottery company provides these statistics is that many people believe that a ball that has not being drawn often recently is 'overdue'. It isn't. There is no connection to link one draw with another. The lottery does not have a memory. We can't use the past here to predict the future. But still we attempt to do so. It is almost impossible to avoid the self-deception that patterns force on us.

Other forecasts are less cut and dried than attempting to predict the results of the lottery. In most systems, whether it's the weather, the behaviour of the stock exchange or sales of wellington boots, the future isn't entirely detached from the past. Here there is a connection that can be explored. We can, to some degree, use data to make meaningful forecasts. But still we need to be careful to understand the limitations of the forecasting process.

Extrapolation and black swans

The easiest way to use data to predict the future is to assume things will stay the same as yesterday. This simplest of methods can work surprisingly well, and requires minimal computing power. I can forecast that the Sun will rise tomorrow morning (or, if you're picky, that the Earth will rotate such that the Sun appears to rise) and the chances are high that I will be right. Eventually my prediction will be wrong, but it is unlikely to be so in the lifetime of anyone reading this book.

Even where we know that using 'more of the same' as a forecasting tool must fail relatively soon, it can deliver for the typical lifetime of a business. As of 2016, Moore's Law, which predicts that the number of transistors in a computer chip will double every one to two years has held true for *over 50 years*. We know it must fail at some point, and have expected failure to happen 'soon' for at least twenty years, but 'more of the same' has done remarkably well. Similarly, throughout most of history there has been inflation. The value of money has fallen. There have been periods of deflation, and times when a redefinition of a unit of currency moves the goalposts, but overall 'the value of money falls' works pretty well as a predictor.

Unfortunately for forecasters, very few systems are this simple. Many, for instance, have cyclic variations. I mentioned at the end of the previous section that wellington boot sales can be predicted from the past. However, to do this effectively we need access to enough data to see trends for those sales throughout the year. We've taken the first step towards big data. It's not enough to say that next week's sales should be the same as last week's, or should grow by a predictable amount. Instead, weather trends will ensure that sales are much higher, for instance, in autumn than they are at the height of summer (notwithstanding a brief surge of sales around the notoriously muddy music festival season).

Take another example – barbecues. Supermarket chain Tesco reckons that a 10°C increase in temperature at the start of summer results in a threefold increase in meat sales as all the barbecue fans go into caveman mode. But a similar temperature increase later in summer, once barbecuing is less of a novelty, does not have the same impact. So a supermarket needs to have both seasonality data and weather data to make a reasonable forecast.

Seasonal effects are just one of the influences reflected in past data that can influence the future. And it is when there are several such 'variables' that forecasting can come unstuck. This is particularly the case if sets of inputs interact with each other; the result can be the kind of mathematically chaotic system where it is impossible to make sensible predictions more than a few days ahead. Take the weather. The overall weather system has so many complex factors, all interacting with each other, that tiny differences in starting conditions can result in huge differences down the line.

For this reason, long-term weather forecasts, however good the data, are fantasies rather than meaningful predictions. When you next see a newspaper headline in June forecasting an 'arctic winter', you can be sure there is no scientific basis for it. By the time we look more than ten days ahead, a general idea of what weather is like in a location at that time of year is a better predictor than any amount of data. And if chaos isn't enough to deal with, there's the matter of black swans.

The term was made famous by Nassim Nicholas Taleb in his book *The Black Swan*, though the concept is much older. As early as 1570, 'black swan' was being used as a metaphor for rarity when a T. Drant wrote 'Captaine Cornelius is a blacke swan in this generation'. What the black swan means in statistical terms is that making a prediction based on incomplete data – and that is nearly always the case in reality – carries the risk of a sudden and unexpected break from the past. This statistical use refers to the fact that Europeans, up to the exploration of Australian fauna, could make the prediction 'all swans are white', and it would hold. But once you've seen an Australian black swan, the entire premise falls apart.

The black swan reflects a difference between two techniques of logic – deduction and induction. Thanks to Sherlock Holmes, we tend to refer to the heart of scientific technique, of which forecasting is a part, as deduction. We gather clues and make a deduction about what has happened. But the process of deduction is based on a complete set of data. If we knew, beyond doubt, that *all* bananas were yellow and were then given a piece of fruit that was purple, we would be able to deduce that this fruit is not a banana. But in the real world, the best we can ever do is to say that all bananas we have encountered are yellow – when ripe. The data is incomplete. Without the availability of deduction, we fall back on induction, which says that it is highly likely that the purple fruit that we have been given is not a banana. And that's how science and forecasting work. They make a best guess based on the available evidence; they don't deduce facts.

In the real world, we hardly ever have complete data; we are always susceptible to black swans. So, for instance, stock markets generally rise over time – until a bubble bursts and they crash. The once massive photographic company Kodak could sensibly forecast sales of photographic film from year to year. Induction led them to believe that, despite ups and downs, the overall trend in a world of growing technology use was upward. But then, the digital camera black swan appeared. Kodak, the first to produce such a camera, initially tried to suppress the technology. But the black swan was unstoppable and the company was doomed, going into protective bankruptcy in 2012. Although a pared-down Kodak still exists, it is unlikely ever to regain its one-time dominance.

The aim of big data is to minimise the risk of a failed forecast by collecting as much data as possible. And, as we will see, this can enable those in control of big data to perform feats that would not have been possible before. But we still need to bear in mind the lesson of weather forecasting.

Meteorological forecasts were the first to embrace big data. The Met Office is the biggest user of supercomputers in the UK, crunching through vast quantities of data each day to produce a collection of forecasts known as an ensemble. These are combined to give the best probability of an outcome in a particular location. And these forecasts are much better than their predecessors. But there is no chance of relying on them every time, or of getting a useful forecast more than ten days out.

We shouldn't underplay the impact of big data, though, because it can remove the dangers of one of the most insidious tools of forecasting, one which attempts to give the effect of having big data with only a small fraction of the data. As we've already discovered, the limitations of sampling are all too clear in the failure of 21st-century political polls.

Sampling, polls and using it all

There was a time when big data could only be handled on infrequent occasions because it took so much time and effort to process. For a census, or a general election, we could ask everyone for their data, but this wasn't possible for anything else with the manual data handling systems available. And so, instead, we developed the concept of the sample. This involved picking out a subset of the population, getting the data from them and extrapolating the findings to the population as a whole.

Let's take a simple example to get a feel for how this works – PLR payment in the UK. PLR (Public Lending

Right) is a mechanism to pay authors when their books are borrowed from libraries. As systems aren't in place to pull together lendings across the country, samples are taken from 36 authorities, covering around a quarter of the libraries in the country. These are then multiplied up to reflect lendings across the country. Clearly some numbers will be inaccurate. If you write a book on Swindon, it will be borrowed far more in Swindon than in Hampshire, the nearest authority surveyed. And there will be plenty of other reasons why a particular set of libraries might not accurately represent borrowings of a particular book. Sampling is better than nothing, but it can't compare with the big data approach, which would take data from every library.

Sampling is not just an approach used for polls and to generate statistics. Think, for example, of medical studies. Very few of these can take in the population as a whole – until recently that would have been inconceivable. Instead, they take a (hopefully) representative sample and check out the impact of a treatment or diet on the people included in that sample. There are two problems with this approach. One is that it is very difficult to isolate the impact of the particular treatment, and the other is that it is very difficult to choose a sample that is representative.

Think of yourself for a moment. Are you a representative sample of the population as a whole? In some ways you may be. You may, for instance, have two legs and two arms, which the majority of people do … but it's not true of everyone. If we use you as a sample, you may represent a large number of people, but if we assume *everyone* else is like you, we can disadvantage those who are different.

In many other respects you are far less representative. By the time we take in your hair colour and weight and gender and ethnic origin and job and socioeconomic group and the place you live, you will have become increasingly unrepresentative. So, to pick a good sample, we need to get together a big enough group of people, in the right proportions, to cover the variations that will influence the outcome of our study or poll.

And this is where the whole thing tends to fall apart. Provided you know what the significant factors are, there are mechanisms to determine the correct sample size to make your group representative. But many medical studies, for example, can only afford to cover a fraction of that number – which is why we often get contradictory studies about, say, the impact of red wine on health. And many surveys and polls fall down both on size and on getting representative groupings. To get around this, pollsters try to correct for differences between the sample and what they believe it *should* be like. So, the numbers in a poll result are not the actual values, but a guesswork correction of those numbers. As an example, here are some of the results from a December 2016 YouGov poll of voting intention taken across 1,667 British adults:

	Voting Intention			
	Con	Lab	Lib Dem	UKIP
Weighted	457	290	116	135
Unweighted	492	304	135	146

The bottom 'unweighted' values are the numbers of

individuals answering a particular way. But the top 'weighted' values are those that were published, producing quite different relationships between the numbers, because these adjustments were deemed necessary to better match the population as a whole. Inevitably, such a weighting process relies on a lot of guesswork.

As a result, political opinion polls since 2010 have got it disastrously wrong. Confidence in polls has never been weaker, reflecting the difficulty pollsters face in weighting for a representative sample. Where before socioeconomic groupings and party politics were sufficient, divisions have been changing, influenced among other things by the impact of globalisation and inequality. It didn't help that polling organisations are almost always based in 'metropolitan elite' locations which reflect one extreme of these new social strata. Add in the unprecedented impact of social media, breaking though the old physical social networks, and it's not surprising that the pollsters' guesswork approximations started to fall apart.

The use of such samples will continue in many circumstances for reasons of cost and convenience, though it would help to build trust if it were made easier for consumers of the results to drill down into the assumptions and weightings. But big data offers the opportunity to avoid the pitfalls of sampling and take input from such a large group that there is far less uncertainty. Traditionally this would have required the paraphernalia of a general election, taking weeks to prepare and collect, even if the voters were willing to go through the process many times a year. But modern systems make it relatively easy to collect some kinds of data even on such a large scale. And increasingly, organisations are making use of this.

Often this means using 'proxies'. The idea is to start with data you *can* collect easily – data, for instance, that can be pulled together without the population actively doing anything. At one time, such observational data was very difficult for statisticians to get their hands on. But then along came the internet. We usually go to a website with a particular task in mind. We go to a search engine to find some information or an online store to buy something, for example. But the owners of those sites can capture far more data than we are aware of sharing. What we search for, how we browse, what the companies know about us because they made it attractive to have an account or to use cookies to avoid retyping information, all come together to provide a rich picture. Allowing this data to be used gives us convenience – but also gives the companies a powerful source of data.

This means that, should they put their mind to it, the owners of a dominant search engine could gather all sorts of information to predict our voting intentions. The clever thing about a big data application like this is that, unlike the knowledge-based systems described above, no one has to tell the system what the rules are. No one would need to work out what is influencing our voting or calculate weightings. By matching vast quantities of data to outcomes, the system could learn over time to provide a surprisingly accurate reflection of the population. Certainly, it would enable a far better prediction than any sampled poll could achieve.

However, impressive though the abilities of big data are, we have to be aware of the dangers of GIGO.

Gratuitous GIGO

When information technology was taking off, GIGO was a popular acronym, standing for 'garbage in, garbage out'. The premise is simple – however good your system, if the data you give it is rubbish, the output will be too. One potential danger of big data is that it isn't big enough. We can indeed use search data to find out things about part of a population – but only the members of the population who use search engines. That excludes a segment of the voting public. And the excluded segment may be part of the upsets in election and referendum forecasts since 2010.

It is also possible, as we shall see with some big data systems that have gone wrong, that without a mechanism to detect garbage and modify the system to work around it, GIGO means that a system will perpetuate error. It begins to function in a world of its own, rather than reflecting the population it is trying to model. For example – as we will see in Chapter 6 – systems designed to measure the effectiveness of teachers based on whether students meet expectations for their academic improvement, with no way of dealing with atypical circumstances, have proved to be impressively ineffective.

It is easy for the builders of predictive big data systems to get a Hari Seldon complex. Seldon is a central character in Isaac Asimov's classic *Foundation* series of science fiction books. In the stories, Hari Seldon assembles a foundation of mathematical experts, who use the 'science' of psychohistory to build models of the future of the galactic empire. Their aim is to minimise the inevitable period of barbarism that collapsed empires have suffered in history. It makes a great drama, but there is no such thing as psychohistory.

No matter how much data we have, we can't predict the future of nations. Like the weather, they are mathematically chaotic systems. There is too much interaction between components of the system to allow for good prediction beyond a close time horizon. And each individual human can be a black swan, providing a very complex system indeed. The makers of big data systems need to be careful not to feel, like Hari Seldon, that their technology enables them to predict human futures with any accuracy – because they will surely fail.

We also need to bear in mind that data is not necessarily a collection of facts. Data can be arbitrary. Think, for instance of a railway timetable. The times at which trains are supposed to arrive at the stations along a route form a collection of data, as does the frequency with which a train arrives at the stated time. But these times and their implications are not the same kind of fact as, say, the colour of the trains. For two years, I caught the train twice a week from Swindon to Bristol. My train left Swindon at 8.01 and arrived at Bristol at 8.45. After a year, Great Western Railway changed the departure time from Swindon to 8.02. Nothing else was altered.

This was the same train from London to Bristol, leaving London and arriving at Bristol at the same times. However, the company realised that, while the train usually arrived in Bristol on time, it was often a little late at Swindon. So, by making the change of timetable from 8.01 to 8.02 they had a significant impact on on-time arrivals at Swindon. The train itself was unchanged. Yet despite this, by making this adjustment, the performance data improved. This underlines the loose relationship between data and fact.

In science, data is usually presented not as specific values but as ranges represented by 'error bars'. Instead of saying a value is 1, you might say '1 ± 0.05 to 99 per cent confidence level'. This means that we expect to find the value in the range between 0.95 and 1.05, 99 times out of 100, but we don't know exactly what the value is. This lack of precision is usually present in data, but it is rarely shown to us. And that means we can read things into the data that just aren't there.

A constantly moving picture

If we get big data right, it doesn't just help overcome the inaccuracy that sampling always imposes, it broadens our view of usable data from the past to include the present, giving us our best possible handle on the near future. This is because, unlike conventional statistical analysis, big data can be constantly updated, coping with trends.

As we have seen, forecasters know about and incorporate seasonality, but big data allows us to deal with much tighter rhythms and variations. We can bring in extra swathes of data and see if they are any help in making short-term forecasts. For example, sales forecasting has for many years taken in the impact of the four seasons and the major holidays. But now it is possible to ask if weather on the day has an impact on sales. And this can be applied not just to purchases of obvious products like sun cream or umbrellas, but equally to sausages or greetings cards. Where it appears there is an impact, we can then react to tomorrow's forecast, ensuring that we are best able to meet demand.

Just as big data allows for more focused seasonality, so too can it provide much tighter regional identity. Retailers in the past might know, for example, that mushy black peas and jellied eels had their own geographic areas of interest. But with big data, every single outlet can be fine-tuned to local preferences.

The pioneers

If we were to look for the pioneers of big data, there are some surprising precursors. In particular, trainspotters and diarists.

Each of these groups took a big data approach in a pre-technology fashion. When I was a trainspotter in my early teens, I had a book containing the number of every engine in the UK. There were no samples here. This book listed everything – and my aim was, at least within some of the categories, to underline every engine in the class. As I progressed in the activity, I went from spotting numbers to recording as much data as I could about rail journeys. Times, speeds and more were all grist to the statistical mill.

At least trainspotters collect numerical data. It might be harder to see how diarists come into this. I'd suggest that diarists are proto-big data collectors because of their ability to collate minutiae that would not otherwise be recorded. A proper diarist, such as Samuel Pepys or Tony Benn, as opposed to someone who occasionally wrote a couple of lines in a Collins pocket diary, captured the small details of life in a way that can immensely useful to someone attempting to recreate the nature of life in a historical period. Data isn't just about numbers.

To transform a very small data activity like keeping a diary into big data only takes organisation. From 1937 to 1949 in the UK, a programme known as Mass Observation did exactly this. A national panel of writers was co-opted to produce diaries, respond to queries and fill out surveys. At the same time, a team of investigators, paid to take part, recorded public activities and conversation, first in Bolton and then nationwide. The output from this activity was collated in over 3,000 reports providing high-level summaries of the broader data. All the data is now publicly available, providing a remarkable resource. A second such project was started in 1981 with a smaller panel of around 450 volunteers feeding information into a database.

However much trainspotters and diarists – and particularly Mass Observation – were precursors of big data operatives, they were inevitably limited by the lack of technology. The best technology I had for my train journey data was a Filofax. If that data could have been pulled together with many other sources into a system, then the essential step from collection to analysis that makes big data worthwhile could take place. It's this kind of process that means that the Mass Observation data is still useful today. And it's arguable that the first step in that direction was a cry of protest from an intensely bored nineteenth-century mathematician.

Doing it by steam – from Babbage and Hollerith to artificial intelligence

The name Charles Babbage is now tightly linked with the computer, and though Babbage never got his technology to work, and his computing engines were only conceptually linked to the actual computers that followed, there is no doubt that Babbage played his part in the move towards making big data practical.

The story has it that Babbage was helping out an old friend, John Herschel, son of the German-born astronomer and musician William Herschel. It was the summer of 1821, and Herschel had asked Babbage to join him in checking a book of astronomical tables before it went to print: row after row of numbers which needed to be accurate – tedious in the extreme. As Babbage worked through the tables, painstakingly examining each value, he is said to have cried out, 'My God, Herschel, how I wish these calculations could be executed by steam!'

Though Babbage could not make it happen practically with his mechanical computing engines (despite spending large quantities of the British government's money), he had the same idea, probably independently, as Herman Hollerith, the American who saved the US census by mechanising its data. Each was inspired by the Jacquard loom.

The Jacquard loom was a Victorian invention that enabled the pattern for a silk weave to be pre-programmed on a set of cards, each with holes punched in it to indicate which colours were in use. Babbage wanted to use such cards in a general-purpose computer, but was unable to complete his intricate design. Hollerith took a step back from the information processor (or 'mill', as Babbage called it), the cleverest part of the design. But he realised that if, for instance, every line of information from the census was punched on to a card, electromechanical devices could sort and collate these cards to answer various queries and start to reap the benefits that big data can provide.

The devices involved were called tabulators. A typical use might be to count how many people there were in different age groups, gender, race and so on. The cards would be passed through the tabulator (initially manually and later automatically), where metal pins completed a circuit by dipping into mercury when they passed through the punched holes. Each electrical impulse advanced a clock-like dial. The operator would then put the card into a specific drawer in a sorting table, as directed by the tabulator (again this part of the process was later automated). Hollerith's tabulators were produced by his Tabulating Machine Company, which morphed into International Business Machines and hence became the one-time giant of information technology, IBM.

The trouble with these mechanical approaches is that they were inevitably limited in processing rate. They enabled a census to be handled in the ten years available between these events – but they couldn't provide flexible analysis and manipulation. One of the reasons that big data can be underestimated is the sheer speed with which we've moved over the last two decades to networked, ultra-high speed technology making true big data operations possible.

When I was at university in the 1970s, we still primarily entered data on punched cards. Admittedly we were using the cards to input programs and data into electronic computers – even so, the term 'Hollerith string' for a row of information on a card was in common usage. Skip forward a couple of decades to 1995, when I attended the Windows 95 launch event in London. In the Q and A at the event, I asked Microsoft what they thought of the internet. The response was that it was a useful academic tool, but not expected to have any significant commercial impact.

By 1995, personal computers, one of the essential requirements for big data, were commonplace, if not yet as portable as smartphones. But the second essential of connectivity through the internet was not envisaged as being important even by as big a player as Microsoft. However, there were already plenty of examples of the third and final piece of the big data puzzle – the algorithm.

Meet Big AI

You can have as much data as you like, with perfect networked ability to collate it from many locations, but of itself this is useless. In fact, it's worse than useless. As humans, we can only deal with relatively small amounts of data at a time; if too much is available we can't cope. To go further, we need help from computer programs, and specifically from algorithms.

Although the *Oxford English Dictionary* insists that the word 'algorithm' is derived from the ancient Greek for number (which looks pretty much like 'arithmetic'), most other sources tell us that algorithm comes, like most 'al' words, from Arabic. Specifically, it seems to be derived from the name of the author of an influential medieval text on mathematics who was known as al-Khwarizmi. Whatever the origin of the word, it refers to a set of procedures and rules that enable us to take data and do something with it. The same set of rules can be applied to different sets of data.

This sounds remarkably like the definition of a computer program, and many programs do implement algorithms – but you don't need a computer to have an algorithm, and a computer program doesn't have to include an algorithm. An example of a simple algorithm is the one used to produce the Fibonacci series. This is the sequence of numbers that goes:

1, 1, 2, 3, 5, 8, 13, 21, 34, 55, 89, 144 …

The series itself is infinitely long, but the algorithm to generate it is very short, and is something like 'Start with two ones, then repeatedly add the last number in the series to the previous one, to produce the next value.'

For big data purposes, algorithms can be much more sophisticated. But like the Fibonacci series algorithm, they consist of rules and procedures that allow a system to analyse or generate data. Here's another simple algorithm: 'Extract from a series of numbers only the odd numbers.' If we apply this algorithm to the Fibonacci series we get

1, 1, 3, 5, 13, 21, 55, 89 …

This data has no value. But if the original data had been about taxpayers, and instead of using an algorithm for odd numbers, we had extracted 'those earning more than £100,000 a year' we have taken a first step to an algorithm to identify high-value tax cheats. I'm not saying that earning more than £100,000 makes you a tax cheat, just that you can't be a high-value tax cheat, and hence worth investigating, if you only earn £12,000 a year. If we were immediately to stigmatise as a cheat everyone selected by the 'extract £100,000+ earners' algorithm, we would be misusing big data. The algorithm is neutral. It doesn't care what the data implies – it just does what we ask. But as users of big data, we have to be careful about our assumptions and know exactly what the algorithm is doing and how we interpret the results.

Once the technology had caught up with the requirements to handle big data, the capabilities of this approach started to make themselves felt. And one of the first areas where big data would have an impact was in an activity which most of us both love and hate. Shopping.

SHOP TILL YOU DROP   3

Good morning, Mrs Smith

For fifteen years I lived in a village with a single post office and shop. It wasn't long after starting to use the post office that the people serving there got to know me by name. Once, I walked into the post office to send off a parcel. 'I'm glad you came in,' said the shopkeeper. 'Last time you came in I overcharged you. Here's your change.' Another time, I went into the shop to buy some curry powder. They hadn't got any. 'I tell you what,' said Lorna, who was behind the counter that day. 'Hang on a moment.' She went through to her kitchen and brought back some garlic and fresh chillies. 'Use these instead,' she said. 'They're far better than curry powder.'

One further example. I used to be an enthusiastic photographer and was a regular at a local camera shop – again, they knew me as an individual. They knew that I brought them frequent business. I had been saving up and decided to take the plunge and switch to a digital camera, so asked what was available in digital cameras for around £400. The answer from the familiar sales assistant was shocking at first. 'I wouldn't sell you a camera in that price range,' he said. I was about to ask him what was wrong with my money when he went on. 'One of the best manufacturers has just dropped the price of its cameras from £650 to £400, but they haven't sent out the stock yet. If you come back in a few days, I can do you a much better camera for £400 than I could today. I really wouldn't recommend buying anything now.'

Look at what he did. He turned away the chance to make an immediate sale. In isolation this is total madness – and it's something that the sales assistants in most chain stores would never do, because they are under huge pressure to move goods today. But this assistant used his *knowledge* of me and the market. He balanced the value of a sale now against my long-term custom. I was very impressed that he had said that he wouldn't sell me a camera now, and that by going back in a few days I could get a much better one. Not only did I go back for that camera, I made many other purchases there. And I passed on this story to other would-be purchasers.

This is what knowledge can do for a local shop. But until big data came along, it wasn't possible to contemplate taking the same approach when running a major chain.

Upscaling

Big data presents the opportunity to provide something close to the personal service of the village shop to millions of customers. As we will see, the approach doesn't always work.

This is partly because of GIGO, partly because of half-hearted implementation – good customer service costs money – and partly because few traditional retailers have built their businesses around data in the way that the new wave of retailers such as Amazon have. But there is no doubt that the opportunity is there.

Such systems were originally called CRM for 'customer relationship management', but now are considered so integral to the business that the best users don't bother giving them a separate label.

The challenge to effective data-driven customer service comes from the way that the data faces two ways: towards the shop (or bank) and towards you, the customer. The shop wants to know as much as it can about the customer, so that it can retain you and get the most money out of you. However, you want the data to enable the shop to give you better service and personal rewards. Done well, big data can provide such a win-win on purchases. And one of the earliest opportunities to do this came in the form of the loyalty card.

Loyalty cards

In my wallet, there are around twenty loyalty cards. Some, typically for hot drinks, are very crude. Here I get a stamp on the card every time I buy a drink, and when I fill the card I get a free cup. It's win-win. I am more likely to return to that coffee shop, so they get more business, and I get a free coffee now and again. However, this approach wastes the opportunity to make use of big data, which is why, a couple of decades ago, supermarkets and petrol stations moved away from their equivalent of the coffee card, giving out reward tokens like Green Shield Stamps, to a different loyalty system. The new card had an inherent tie to data and, in principle, gave them the opportunity to know their customer and provide personalised service like the village shop.

With my Nectar card, or Tesco Clubcard – or whatever the loyalty programme is – I no longer have a stamp card, where all the data resides in my wallet. Now, every time I make a purchase, I swipe my card. From my viewpoint, this gives me similar benefits to the coffee card, accumulating points which I can then spend. But from the shop's viewpoint, it can link me with my purchases. The company's data experts can find out where and when I shopped. What kind of things I liked. And they can make use of that data, both to plan stock levels and to make personalised offers to me, for example when a new product I might like becomes available. The system simulates the friendly local shopkeeper's knowledge of me, making it possible for me to feel known and appreciated. The store gives me something extra, which I as a customer find beneficial. But this kind of system can't work effectively without deploying big data.

Loyalty cards got over the anonymity of cash. As it happens, though, such cards are probably coming to the end of their useful life. This is because we are paying for things less and less frequently with traditional forms of payment such as cash and cheques. If we use a debit or credit card, the shop can make exactly the same kind of big data connection as it would with a loyalty card. This is a move that has been on the way for at least twenty years.

Mondex and more

When I first moved to Swindon in the mid-1990s it was to discover the tail end of a revolutionary experiment called Mondex. Most Swindon shops had been issued with card readers for the Mondex smartcard. The card could be loaded with cash at cashpoints and also via a special phone at home. The idea was to trial the cashless society. Whether because it was only months before the experiment ended, or because it had never proved hugely popular, I found many retailers were surprised by my requests to use Mondex. But it certainly had its plus side.

There was no need to carry a pocketful of cash – and the ability to add money to the card at home made visits to the cash machine a thing of the past. But in the end, the smartcard was trying too hard to emulate the physical nature of money. We used to carry cash around – now we carried virtual cash on a card. It was easier, but it replicated an unnecessary complication. It was a small data solution – the Mondex card knew very little about me and had no connectivity – in an increasingly big data world.

Instead of going down the smartcard wallet route, the new wave of cashless payments makes intimate use of big data. 'Tap and pay' card systems do away with the need to load up a smartcard (a failing that is even more obviously a pain with London's dated Oyster card, with which any online payments have to be validated at a designated station), because the new contactless cards are simply a means of identification linking to a central, big data banking system. Even better from a security viewpoint are phone-based payment systems such as Apple Pay, with the same convenience, but the added security of fingerprint recognition.

Arguably this new generation of cashless payment is itself transitional. As it stands, banks need layers of security to handle fraudulent use of cards, security that is driven by big data. Like me, you may have received an automated call from your credit card provider, asking you to confirm that you had made the last three transactions, as you were buying in a pattern than was not normal for you. Whenever it has happened to me it was simply that personal circumstances were unusual – for example, when my daughters were starting at university and we had to buy all sorts of household goods. But in some cases, these systems prevent fraudulent use. It reflects the ease of separating us and our cards. If we could move away from paying with cards or our phones, then there would be less fraud.

All we use the card or phone for is to link an individual to a bank account. With the card it's via a PIN (not even that for contactless payments) and with the phone the link relies on the phone's fingerprint reader. But it's hard to imagine that payment systems won't eventually use biometrics such as fingerprint or facial recognition directly. Admittedly, as thrillers occasionally demonstrate, there are still ways to fool biometrics by copying or removing body parts, but with effective biometric recognition, big data can duplicate the friendly local shopkeeper who knows you on sight.

London's early attempt at an electronic wallet, the Oyster card, now looks to have had its day. It has always been somewhat flawed. Not only is it less flexible than Mondex, as you can't load it up directly at home, if you elect to pay online, you have to specify which station you will use to get the 'cash' on to the card, at least 24 hours ahead of doing so. But Transport for London has signed the Oyster's death warrant by accepting contactless smartphones, credit and debit cards. There will remain a niche market for electronic wallets like Oyster for, for example, children – but for most of us, a night out in London can now be undertaken with a single card or phone. Let's take such a journey with an enabled smartphone and see how big data works in two distinct directions.

A night out in London

I summon an Uber taxi with my phone. The details of my journey, the identity of my driver, and our respective ratings are added to Uber's information on me. Uber also links to my bank, accesses funds and provides them with information on the payment and the broad location of the transaction.

One of my friends finishes work earlier than the others so I'm meeting him first. As I arrive outside his office, my phone flags up that there's a Starbucks around the corner, so I send him a message to meet me there. I place an order and pay on the phone using my Starbucks app before I get there, so that by the time I arrive in the coffee shop, my skinny latte is waiting on the bar. Here, the whole transaction was made with Starbucks. The company knows when I was in which shop and what I ordered (and awards me an electronic loyalty star). It's only when I need to top up my Starbucks account that I also get an interaction with my bank account, which won't know about my visit. (The Starbucks card is an interesting compromise. Although, like the Oyster card, it has the inconvenience of having to be preloaded, this process is made easy – I can do it with a couple of taps and a finger-print from my phone – and, as mentioned, I'm provided with loyalty rewards for using it.)

My friend joins me and we drink our coffee. Then we check where the others are up to: we've arranged to meet at a restaurant but a couple thought they might be delayed at work, so we've added each other to the 'Find my Friends' app. I glance at this and see that one has reached the restau-rant, one is five minutes' walk away from the venue and the third is still at work. Probably time to set off. Exactly how much information the app provider keeps in this case is not entirely clear, but in principle it could know where we've been ever since we activated the app.

I check the location of the nearest tube station on Google Maps. To make it convenient, I've connected this to my Google account, so again, in principle, data about my move-ments could be stored by Google. At the tube station, I use my phone to tap my way through the barriers. This is the most complex transaction of the night. Apple's systems now have the chance to note where I am and what I am doing. The cost of the ticket also goes to Apple, which will then dispatch a payment request to my bank. The bank then sends me a text alert to confirm that I have used Apple Pay for this process. And the payment goes from the bank to Transport for London via Apple. Three different major systems have found out more about me and my actions.

Finally, I go to the restaurant where I can pay my bill via PayPal or Apple Pay from my phone – and again I'm potentially providing the restaurant, the bank and Apple with knowledge of me and my actions. All of this is today's technology. From my viewpoint, I have had an effortless evening. I didn't need to carry cash and my payments were secure thanks to the technologies employed. Big data made my life more convenient. In exchange, the companies concerned found out more about me, about my likes and my activities.

The big data component of my night out could be even more advantageous from my viewpoint – the various companies could, for example, offer me discounts on my favourite purchases or activities. But it is also a huge benefit for the companies. They are amassing data that helps them to know me better and sell to me more frequently or encourage me to spend more. And there is also the concern of misuse. It's easy to say 'If you've nothing to hide, why does it matter?', but being under surveillance by companies who may not have your best interests at heart could be a serious concern. Big Brother may be hiding in the data, as we shall discover. But, as yet, it won't stop me using these technologies – because big data ensures that they work together to make my life easier.

In this section I have described a night out. A day's shopping would undoubtedly provide similar examples – but increasingly, big data is ensuring that I don't need to leave home to shop, if the act of shopping doesn't interest me.

Comparison shopping and Amazon scanners

Much of the modern shopping experience is driven by data: discovering what's available, choosing the product that works best for you and getting the best price. But the high street isn't necessarily the best place to acquire this data. As you walk down the street, each shop largely controls its own data – but big data and connectivity is breaking down that control.

One big breakthrough is comparison shopping. Once you know what you want, you inevitably want the best price. You might try haggling in the shop – and surprisingly, quite a few shops, even big chain stores, will do this if you can speak to a manager. However, it would be very convenient if you could send a minion around a shop's competitors and see if you can get the product cheaper elsewhere. Online, this process becomes much easier. To sell online, shops have to open their data up to a degree, and this means that a comparison shopping site can bring together prices from different retailers and give you a heads-up.

However, comparison shopping sites can be expensive to set up and run. The first such site in the UK was a comparison site for books, called BookBrain, and though there are still several attempting this service, they remain small. On the whole, when comparison sites are successful, they have limited themselves to markets where commissions are high, such as insurance and travel. And commissions are an important factor to consider.

Comparison sites don't sell anything themselves, so to make money they have to get a fee from the retailer they send you to. This is why you don't see comparison sites for some areas of the market. Of course, commission also opens up the possibilities of being misled, or being provided with incomplete information. It would be in the site's interest to point customers towards retailers providing the highest commission, whether or not they are the best match – and there is no point in a site including retailers who don't give commission.

Even so, a comparison site is a weapon of big data that is largely beneficial to the customer who hasn't time to search through the different possibilities. But such a site is useless if you don't know what you want to buy in the first place. No matter how sophisticated the website, the browsing experience when exploring without a clear purchase in mind fails to match up to walking round a conventional shop. The variety of stock that a virtual store can offer becomes a disadvantage – there is simply too much to look at. For some kinds of browsing, you can't beat being in a physical store. Even so, big data tries to play its part.

Walk around a large store these days and you will sometimes see people pointing their phones at products. Some may just be taking a picture of a product to remember it or ask someone else about it. But others will be scanning the barcode into a retail app like Amazon's to see if they could get the product cheaper online. The bricks and mortar store is landed with all the cost of providing the space to browse, while the online retailer gets the sale. In such circumstances, you've won on price and the online retailer has won via big data. The only problem is, if you drive the bricks and mortar store out of business by taking advantage of it this way, you have lost your opportunity to browse, which may be why Amazon, for example, is now starting to open physical stores.

More often than not, big data gives convenience to the

user, but a stronger benefit to the business. However, done properly, the big data approach can offer the human part of the equation a seamless experience that makes it seem almost magical.

Training the system

I recently travelled by Eurostar train from London to Brussels. Someone else had booked the tickets for me and sent an email with the booking. That email had a link to click, which brought up a web page. This also had a link to click. A couple of seconds later, there was a ping in my pocket, and my tickets appeared in the wallet on my phone. All I had done was click twice with the mouse. I entered no information whatsoever – though this seamless effect only worked because I was using a web browser on my computer which was linked via Apple's cloud service to my phone.

I now had on the phone the timetable for my journeys and a barcode to scan at the ticket barrier. It was almost all I needed for the journey, though things could have been even better. I still had to mess around with a paper passport (despite the modern passport containing a chip with biometric data). This now felt like a transitional item in the big data world. But the ease with which someone could email me details and two clicks later I had tickets on my phone was big data at its best. It benefited Eurostar, but I got a lot from it as well. However, not all big data actions in sales and marketing feel quite so beneficial. Some are very one-sided.

Shopbots and Botshoppers

One step away from applications that use data to help the customer sit mechanisms which enable big data to become the customer, forming a kind of self-fuelled market. There are many examples of these, but let's look at two: Amazon Seller shopbots and stock market AIs.

On Amazon, you can buy many products from 'other sellers'. Amazon acts as a marketplace, taking a cut of the price in exchange for making the sale possible – but the price is set by the seller. This makes it tempting for the seller to do two things: to tweak the price up if no one is undercutting them, and to tweak the price down if they aren't the cheapest seller.

Large sellers have thousands of lines listed, making it impractical for a human to cope with all the data involved. Enter the shopbot – an algorithm which, on a regular basis, will make those tweaks in pricing. In principle, this means that the seller is kept in the best position. But if those rules aren't carefully restricted – or the frequency with which the shopbot checks pricing is too high – the big data approach can get out of control. Excessive tweaking down results in far too many products listed at 1p – even when it isn't financially viable to post them at this price.

Similarly, where the bot is set to tweak up if there is no opposition, you will find, say, cheap and cheerful second-hand books priced at hundreds of pounds a copy. We're not talking first editions of classics here – but perhaps a 1993 *Newnes MS-DOS 6.0 Pocket Book*. As it happens, I have a copy of that title and just checked on Amazon Marketplace to find that the only new copy listed had suffered this kind of shopbot tweaking. It was priced at £999.11 – so I listed my copy at £999.10. (If you take a look now, you may find that the shopbot has tweaked the other seller's price down, so my £999.10 is now the highest.) It's fine to ask a premium for a rare book, even if it is worthless, because someone (the author, say) may eventually want a copy. You may remember the advertisement for Yellow Pages featuring *Fly Fishing* by J.R. Hartley. However, there are limits. The big data approach is only as good as the data and the algorithms that handle it.

The stock market example is the reverse of a shopbot. Here, the bot *replaces* the shopper. You can see in principle why someone might think it's a good idea to make use of big data to purchase or sell the right stock at the right time. But what we're doing here is the equivalent of having a comparison shopping site where the customer is given no information at all and the site simply buys the insurance (say) that it thinks is best for you. That's a scary prospect, as stock markets discovered to their cost with the flash crash.

It was 6 May 2010. In 36 minutes, starting at 2.32 in the afternoon, New York time, over a trillion dollars was wiped off the value of US stocks. To blame were a collection of botshoppers. Many stock market trading deals are no longer handled by human beings. Instead, algorithms react to stock market data. In this case, the algorithms in question – what I'm calling botshoppers – primarily belonged to high-frequency traders, who often only hang on to stocks for minutes before reselling. Some of the algorithms were badly written. This made it possible to get into a spiral, where sales during one minute were based on a percentage of sales in the previous minute.

Big data systems can act far quicker than human beings, which is why in that 36 minutes that it took humans to work out what was happening and to pull the plug on the algorithms all hell broke loose. This is a very costly example of the rule that big data is only as good as the algorithms that handle it. Make a mistake, and the high speed and repetitive ability of the algorithm means that it can do a lot of damage before the effects are noticed.

Shopbots and botshoppers are at least doing a job that can be useful. But there are those who would argue that a common application of big data to shopping is not just useless but downright dangerous. This is when big data gets control of advertising.

We know what you're looking at

It's creepy the first time it happens. Arguably, it's creepy *every* time it happens. Let me give a specific example. A few days ago, I had a conversation with my wife about a present we were buying one of our daughters for Christmas. Today, when I opened Facebook, I was presented with an advert for that specific product. (A Chanel perfume, if you must know. Our daughters have expensive tastes.) This advert genuinely shocked me, because I hadn't even looked for the product online. Had Facebook been listening in on our conversation? There are at least two technologies, which we'll come on to later, that make this theoretically possible.

As it happens, there was a simple and less intrusive explanation for the extreme spookiness. My wife had used my computer to compare prices on this product when I was out. The trail left by her searches – probably a cookie, the small file used by a website to hold data from visit to visit – had been picked up in Facebook to provide a 'targeted advert'. But even that level of intrusion is too much for most of us. According to a US survey, 68 per cent of respondents disapprove of targeted advertising because of the feeling that they are being snooped on – the percentage goes up for those with college degrees and higher annual incomes – despite this, such adverts have become a feature of our lives, driven by big data.

Here a benefit of big data to the retailer is being forced on us. In theory, it's good for us as well, the argument being that we would rather see adverts for products we are interested in than for something we hate. But there's a balance between usefulness and intrusion – and most seem to feel that, in this case, the intrusion outweighs the benefit.

There's more to it than that, though. If you've ever been to a travel website, for example, and seen advertisements for competitors you might think this is targeted advertising that has gone wrong. Maybe Google has fed them the wrong ads, as surely a travel site wouldn't advertise its competitors? You wouldn't go into Sainsbury's and see adverts for Tesco and Waitrose. However, if you do see adverts for competitors online, the chances are that big data has identified you as a timewaster and is converting you into a revenue stream.

Anyone who has worked for a retailer knows that after a while you can spot some individuals who are very unlikely to make a purchase. When they take up a lot of a sales assistant's time, they are literally timewasters, as the store might lose other business because the sales assistant is tied up. This doesn't happen online (unless the site is so busy that it becomes unbearably slow). But algorithms can still try to spot the online equivalent of the timewaster for financial gain.

If the system decides that you are very unlikely to make a purchase, that's the perfect time to allow a competitor's advert to appear. If you click on the ad, the site you were visiting gets a small commission – better than nothing at all, and better still, their competitor pays for it. Of course, this is only a good move if the system can read your mind to discover that you won't make a purchase. It tries to do this by combining obvious signals, such as whether or not you log into the site and whether visitors from your IP address have made purchases before, along with any time of day, time of year and search topic data that suggest the probability you'll make a purchase. It's a guess – but if it's likely to be right more often than not, then it's easy enough to set a level at which errors become acceptable to the company.

What this doesn't determine, though, is whether or not an error in targeting will prove irritating to the customer. Let's go back to more general targeted advertising. It might be very clever that the system spots you've been looking at a specific perfume and offers you that product. But it ceases to be clever and becomes frustrating, even if you like targeted advertising, when the ad comes up after you've bought the product (especially if you've just spent more on it). In other cases, the targeting can be hugely misjudged. Comedian and statistician Timandra Harkness tells the story of an interviewee who had flown down to Florida to take care of her ageing mother who was ill. The next day, she started getting targeted advertising for funeral services in the area.

However, even such an offensive miss is arguably better than targeted advertising that intentionally preys on the vulnerable. It does not seem right, for example, that someone who is looking for information on debt management should then start to receive a stream of adverts for payday loans – infamously a burden to those with debt problems due to their four-figure APIs – yet this regularly happens. It's not enough to weigh up the benefits of using big data this way, we also need to assess the human costs. It's not clear-cut. It may be a good idea, for example, that people who are struggling with debt receive targeted advertising for free, helpful debt management charities. But one thing is for certain: it's a moral decision that we can't leave to an algorithm.

To select an appropriate ad, the targeting algorithms keep an eye on your browsing history. But that's only possible online. Surely this could never happen in the real world?

We're watching you

It's hard not to feel sorry for bricks and mortar retailers. Not only do online stores have far lower overheads, they can access all kinds of information about shoppers that isn't available on the high street. At least it isn't available today – but it's very close. We're starting to see a little targeted advertising, whether it's the TripAdvisor app popping up and telling you about the handy restaurant round the corner, or display adverts that can detect your mobile phone and attempt to display something relevant to you. But far more is happening inside some stores in the developing field of video analytics.

If you've ever watched a crime drama on TV, the detectives are often landed with looking through hour upon hour of CCTV footage to try to spot an individual or a car. It's easy to think of big data as being just about numbers – but a video is just as much data as a spreadsheet, particularly in a digital form where it is simply a collection of ones and zeros. Humans are very good at processing visual data. But this ability tails off with time. We aren't good at watching many hours of video, looking out for a specific trigger. This makes video analysis an ideal opportunity for big data algorithms to step in. It's something increasingly used in detection – but also in retail.

No one is surprised to see a video camera in a store. We'd be more surprised not to see them. But the assumption is that they are security videos. With big data, though, they have the potential to be far more. Cameras can track which areas of the store, and which displays, get most attention. Individuals can be flagged up using facial recognition, not only to track their activity in the store but also to identify repeat visits. It wouldn't be too much of a stretch to extend that facial recognition to social media – at which point it would be possible to identify individuals who have visited your stores, and contact them online with offers.

Another approach, taken by US company Percolata – a provider of systems used by retailers including Uniqlo and 7-Eleven – combines information on customer flows into and around the store with the sales that each employee rings up. This is an attempt to come up with a measure that is difficult to assess in customer service – productivity. The system monitors how much each sales assistant takes, per customer going through the store. This irons out the variability for quiet and busy times – but it is still a measure that suggests the only valuable thing a sales assistant can do is take money.

Such data, plus various combinatorial factors – the impact of different sales assistants working together on takings, the people who work better on sunny or rainy days – is then used to drive a system that allocates shop floor time to the assistants at short notice to maximise the revenue of the store. (See the section on the 'gig economy', page 115, for more on this kind of system.) By comparing similar stores with and without the system in use, Percolata suggests that they can push up revenue by between 10 and 30 per cent. Like many algorithm-managed jobs, this is not likely to be a positive experience for the employees. But what about the shoppers? Is Percolata's system, or making use of video analysis, intrusive or straightforward commercial activity?

Leaving aside the possibility of linking facial recognition to social media, it's hard to draw the line. Few would have a problem with a system that monitored footfall around a store anonymously, but once individuals are identified and tracked, this becomes closer to surveillance. As shoppers, the chances are we would never know that this was happening. But employees could also be monitored, giving stores the opportunity to reward good sales technique – and punish shirking off. Good management practice or intrusive behaviour? They are grey areas.

Such big data integration with video is not limited, of course, to shops. The police and civic authorities could use it via CCTV on the streets. Transport hubs, such as airports, particularly those where security is a serious issue, are already experimenting with facial recognition on videos to track individuals and flag up suspicious characters. However, this isn't the only place that big data is at work at the airport, something you will have experienced if you have ever turned up to find that your flight is overbooked.

Why airlines overbook

It seems crazy. Airline booking systems were among the earliest effective users of big data. Within the industry, American Airlines used to be so focused on its Sabre booking system that it described itself to fellow airlines as a booking system company that also flew planes. With split-second, real-time booking happening all over the world, few companies have a better instant notification of the state of sales and inventory than an airline. And yet for years now, airlines have significantly overbooked some of their flights. This is not an accident. It is a commercial decision to take a calculated risk.

The discipline operational research (OR), in which I used to work at British Airways, was responsible for this overbooking technology. OR began during the Second World War, when physicists and mathematicians were brought in to help with military problem solving. They produced mathematical mechanisms to, for instance, decide what pattern of depth charges was most likely to put a submarine out of action. But after the war, this expertise was dispersed into industry. And OR analysts are masters of the algorithm.

Intentional overbooking was a direct result of the flexibility of business travel. A full price ticket was very expensive, but was fully refundable, even after the flight, if it wasn't used. So, when business travellers had uncertain schedules, they would buy several tickets around the time they hoped to travel. They would only use one, and get a refund on the rest. This meant that some flights took off with lots of empty seats, which had nominally been sold, but would later be refunded.

This behaviour was very route dependent. It was much more likely to happen on, say, Heathrow to New York or Heathrow to Amsterdam than Luton to Marbella. OR teams began to collect data on the statistical distribution of the no-shows. And as a big data picture built up, they could predict with reasonable accuracy that, say, 10 per cent of bookings on a particular flight would not be used. And so, the systems began to sell 110 per cent of the seats on that journey.

Like all predictions of the future that are based on the past, such 'overbooking profiles' were not foolproof. Often they were effective, but occasionally more passengers would turn up than there were seats on the plane. The airline would then have to pay off some passengers to fly at another time, compensating them for the inconvenience. Some passengers without a strict timetable came to enjoy this source of easy money. For the airline, there was a simple balance of cost and benefit. Although compensating passengers who were bumped off their flight had a cost, it was more than outweighed by all the extra revenue from being able to sell more tickets.

And so the apparent misuse of data in overbooking has become a way of life in an industry with a great ability to make use of data. Although overbooking makes it appear that airlines are getting big data wrong, in reality big data brings them a hidden reward. It's harder, though, to understand the way that high street banks deal with big data.

The cautious banker

As we have seen, big data has had a huge role in stock markets and other banking activities that are closer to gambling than careful handling of money. But everyday high street banking – undertaking the activities that deal with our bank accounts – seems to have a poor grasp of the importance of big data. Like the airlines, the banks were early into large-scale, real-time transaction processing – the ideal environment to make big data work for them and their customers – but the outcome has been very mixed.

It's true that banks make use of big data in a number of ways that impact customers. Where once it was your friendly(ish) local manager, who knew you personally, who would decide whether or not to lend you money, now it's down to the application of an algorithm to a data collection that extends outside your account activity to credit scoring agencies and further afield. This makes for quick decisions, but, as we'll discover in Chapter 6, it can also have a terrible impact on everyday lives when a faceless algorithm decides your future.

Similarly, as we have seen, big data systems constantly monitor credit and debit card activity for fraudulent activity.

It's obvious how this kind of system is as much a benefit to the customer as it is to the bank. But what is less obvious in banks' data handling is why they have archaic time lags in their systems.

You might have, wondered, for example, why some payments take several days to clear – or why a monthly standing order that usually goes out regularly as clockwork will be two days late if the date falls on a Saturday, or even three or four days late if bank holidays intercede. This is because banking systems were built to simulate the old paper-based systems, where such delays were necessary to enable physical pieces of paper to be manually checked and passed from branch to branch through internal mail systems.

Rather than rewrite the systems from scratch – which would have huge overheads – the banks have bolted on new capabilities. So, we can transfer money electronically from one account to another in seconds – even at weekends. We can pay instantly 24/7 with a tap of a contactless card. But these activities sit on top of a system that still thinks it is dealing with paper ledgers and cheques that have to be returned to the branch of the customer who wrote them before they are cleared. The banks demonstrate well that big data in everyday life is in a transitional phase. It is only when we have a new generation of systems built with a modern approach to big data in mind that everything will catch up.

As customers, we tend to prefer our banks to be conservative. We would rather they were careful with our money than carefree. However, big data does manage to embrace fun as well as the serious side of commerce. And that's what we will discover in the next chapter.

FUN TIMES

4

Visiting the Australian garden

Big data – and big data fun – has come into our lives most directly through the internet, and specifically the World Wide Web. The two are often confused. The media frequently claim that Tim Berners Lee, the British computer scientist working at CERN in Geneva, invented the internet. He didn't – his contribution was the Web.

The internet is an infrastructure for connecting computers. It's literally an 'inter-computer network'. It developed organically in the 1970s, growing out of a US military network called ARPANet, which from its 1960s beginning had become a significant presence in US universities. Initially, the network was used for logging remote terminals on to a distant mainframe – so a researcher in, say, Los Angeles, could interact with a computer in Boston without travelling.

The first big step away from such basic interconnectivity came in September 1973 when someone forgot his razor.

This was US computer scientist Len Kleinrock, who had been at a conference at Sussex University in the UK and returned home to Los Angeles, only to discover he'd left his razor in Brighton. As the Brighton conference was on networks like ARPANet, a temporary extension to the network had been set up, relaying its signals via the Goonhilly Downs satellite communication station in Cornwall which usually handled transatlantic telephone calls and TV.

The link was still up when Kleinrock got home, as some of the delegates at the conference had stayed on. He discovered a colleague connected to the network (despite it being 3am in the UK) and sent a message via a program designed for connecting teletype devices. Kleinrock's request to retrieve his razor was effectively the first email.

Through the 70s, email spread, to be joined by message boards and other communications mechanisms using either the internet or commercial networks like CompuServe and AOL. But big data was still to get a toehold in everyday life outside the enthusiasts until Berners Lee worked his magic.

Though the name 'World Wide Web' sounds grandiose, all Berners Lee attempted was to make it easy to access a library of electronic documents via the internet. He set up a standard mechanism for this, just as the internet founders had devised communication protocols enabling inter-computer communication to function. And Berners Lee made use of the concept of clickable hyperlinks to jump from document to document, introduced conceptually by Ted Nelson in the 1960s and already widely used in Microsoft's help systems and the Mac Hypercard program. Berners Lee got the first local web functionality together at CERN in late 1990 and opened it to the world in 1991.

When I first played with the web in 1992, there were only a few websites. Like CERN's, these were primarily text documents, with very little in the way of images – not surprising, bearing in mind that outside of major organisations, access was by dialling up and interacting through a modem, around 1,000 times slower than a basic modern internet connection. There was no Google, nor any other search engine (the first big search engine, AltaVista, came along in 1995). You had to know the specific address to type into the crude web browser (which for no good reason had a hard-on-the-eyes grey background). Probably the most exciting website to visit was the Australian Botanic Gardens site, started in 1992. Mostly text, this had a few low-resolution images. But the thrill was that you were looking at material that was coming directly from Australia. The world was suddenly much smaller.

It would have been impossible to believe back then just how much the web would change everyday lives, particularly as a universal source of information, pumping big data to our fingertips.

The answer to everything

You are watching TV and someone gets mentioned on the news. 'Who was he married to?' someone asks. There was a time when finding this out would have been almost impossible without heading down to the library – and even then, the chances are that the information would not be available. You could, perhaps, find it after a day or two running through celebrity columns in archives of newspapers. In practice, though, you wouldn't bother to retrieve such trivia. Now, it's a matter of picking up a smartphone, typing in a few words and the information is there. It's hard not to see the internet as a universal big data oracle, the source of all the information you might need, where and when you need it.

Of course, it's not that simple. The internet is not an encyclopaedia. Much of the information that turns up in response to a web search is not curated – so judgement has to be made about what is factual and what is fabricated. There was a lot of concern in 2016 about 'fake news sites' and their influence on the US presidential election – when information is easy to publish and easy to reach, we need considerably more discipline about checking information before taking it as gospel.

It's the classic Wikipedia problem. There has never been an encyclopaedia before with the tiniest fraction of the information that there is in Wikipedia. And a remarkable amount of that information is accurate, particularly on science and technology topics. Analysis in the past has shown that there were no more errors in science articles of Wikipedia than there were in *Encyclopaedia Britannica*, but Wikipedia contained far, far more information. Where articles stray into more contentious topics, though – politics for example – it can be hard not to discover oddities creeping in, when you have a system that is only lightly policed, despite Wikipedia supporters' efforts to keep on top of things.

Sometimes misleading information can be inserted for

fun. For some time, the Wikipedia entry on the Surrey children's attraction Bocketts Farm contained this information:

> Bocketts Farm is also one of the world's first complexes successful in genetically engineering dinosaurs, the first of that being an 18 tonne bronchoraus [sic] named Stuart, who grazes a 16 acre paddock upon the north end of the Farm Park. His diet comprises primarily of hay, vodka martinis and flying saucers. In the future, Stuart says he would like to pursue a career in accountancy.

Needless to say, this is not entirely accurate. Nevertheless, the internet has provided a data revolution in serving up information, an environment to which society is yet fully to adjust. The key to such information access is a good search engine, a market that – with an honourable mention going to Microsoft's Bing – is dominated by one name: Google.

Google it

There is something magical about the Google search engine. If I enter 'Brian Clegg', within a fraction of a second Google claims to have found about 564,000 results. And though some are for an arts and crafts supplier of the same name, a surprisingly high number cover exactly what I'm looking for. This is big data at its most impressive. Even with the remarkable speed of modern technology, it would be impossible to look through each page that I can potentially access – Google covers between 47 and 49 billion pages (by comparison, Bing covers 16 to 17 billion, though realistically most of the missing pages will rarely be used).

That doesn't mean, incidentally, that there are only 50 billion pages on the web. There are plenty more documents that Google is prevented from looking at because they are part of a commercial or secure site. However, the range of pages Google covers is still vast enough to make it hard to believe that it can really be searching through all this material for us. Of course, the site doesn't work its way through the entire web each time we make a request. Its software agents, called crawlers, constantly roam the web looking for new material to add to its indexes, and it is these that our requests are matched against, rather than the raw data. Even so, the Google index runs to 100 petabytes of data, where a petabyte is a million gigabytes.

Finding responses from the index is partly about matching the words in your query to the words on web pages – something that is speeded up by starting the process as soon as you begin to type – but it's far more. Google uses a whole mix of data to put the results in a certain order. Some will get to the top because their site owners have paid for them to do so. Others will bubble up the order because they are recent, because the site is linked to by other important sites, because Google considers the site to be high quality and more. The results will even be structured differently depending on anything Google's system can deduce about the person who has requested them from browsing history or being logged in to Google's wider facilities. A whole industry has built up trying to reverse engineer the secret sauce that is Google's ranking algorithm and to push sites up the ranking.

To counter such 'search engine optimisation', Google's engineers are constantly tweaking the ranking algorithm.

Of course, Google can be used for far more than entertainment. Many searches are undertaken to buy something, or for business or educational purposes. But there's no doubt that the fun side of big data is at play on the internet too, and never more so than with the monsters of video streaming, Netflix and Amazon Prime.

Netflix and chill

If we set aside online gaming, where the big data aspects are evident, the most obvious entertainment focus of the internet is watching videos – and as we have seen with Netflix's development process for new series, there's far more involved here than simply getting the bits and bytes of digital moving pictures from a server somewhere in the world on to your TV, laptop or smartphone. Even so, the apparently simple ability to click on a movie or TV show of your choice at any time, starting and stopping it as you would a DVD, is phenomenal. A single DVD holds several gigabytes of data. Imagine the sheer quantities of data involved when these streaming sites are serving millions of customers. This is *industrially* big data.

Although streaming services have yet to become as widespread as conventional broadcasting – at the time of writing, around a quarter of UK homes subscribe to the brand leader, Netflix – they are beginning to change the way that we watch television. Increasingly we expect to watch what we want when we want, and as the on-demand services build their customer base, they are able to fund sufficient new material that it is possible to do all your viewing through these non-conventional data channels.

It is hard to believe that in twenty years' time there will be scheduled broadcasts any more, with the exception of live events. Even the traditional broadcasters, such the BBC and ITV in the UK and CBS, NBC, ABC and Fox in the US, are unlikely to bother with an increasingly outdated way to make TV accessible. In the future, we can expect all the main networks to be on-demand. As Netflix has shown, with that comes some significant benefits for the network – far more than saving money by losing the overheads of broadcasting or providing a dedicated cable network. It's easy for nostalgia to give us the idea that we would lose out by moving away from traditional broadcasting. But all the evidence is that it would enable traditional broadcasters to – like Netflix – take more daring and effective decisions in their commissioning of new programmes.

Like all 'pure data' entertainment, video has the potential to be pirated. There will always be some piracy, but the evidence is that the best way to minimise this is to make legal streaming or downloaded entertainment as easy, convenient and supportive as possible. Companies like Netflix managed this well from day one, making their product available through as many viewing platforms as possible and providing valuable features, such as being able to stop watching a movie or show part way through and restarting at the same point on another device.

This has typically been the difference between big data leaders like Netflix and Amazon Prime and digital access to video provided by the old studios and networks. The old guard has typically made it relatively inconvenient to stream – because they want to boost direct viewing to support advertising revenue and DVD sales. Consequently they have suffered more than the new kids on the block at the hands of the pirates.

However, every kind of TV and film company has to make moving pictures in the first place – and this is another part of the process where big data is showing its hand.

Fixing the picture

A quite different application of big data to TV and movies is in making images in the first place. Remarkably, an artificial intelligence system has been trained to produce moving images as a kind of 'what happens next?' puzzle from a still picture.

A team from the Massachusetts Institute of Technology plugged into the system the big data of 2 million videos from an online sharing site, only selecting videos that featured a specific set of scenes, including babies in hospitals, beaches and railway stations. The AI system made use of this big data set to generate video sequences from a still image. As of 2016 these are short, around one second long, but are nonetheless managing to animate a single, still photograph.

As always with big data, the outcome is only as good as the algorithms making use of the data. Cleverly, the MIT system used two separate algorithms, one of which acted as a critic for the quality of the output of the other, looking for variations from the expectations *it* had from all the movies *it* had absorbed.

These systems are inevitably limited. We make deductions (or, rather, inductions) based on wider knowledge. If we see a talking human head sticking out of the sand on a beach, we assume that the rest of a person is buried in the sand. The system could only realise this if it had processed a video of someone being buried. And at the moment they are limited in resolution and in how far they can go. In principle, they could be used to fill in a few missing frames in a movie. But the real reason the researchers are trying to do this is to give their artificial intelligence systems a better grasp of 'what happens next' – something that is essential when we want technology to operate with autonomy. Take the example of self-driving cars, already being tested on the roads. In deciding how to act, the artificial intelligence system controlling the car has to monitor the environment around it and predict outcomes to reduce the risk of accidents. The MIT system could be a step on the way to improving this ability.

Video is not the only part of the entertainment industry that has been making the difficult transition into the big data world. But music and book publishing have been more like the old studios in their response to the impact of big data: they have struggled with new ways of working.

Moving the media

The digital world had an impact on music first, where piracy proved to be an immense challenge. By its very nature, digital data is easier to copy than physically stored analogue data. Just moving from LPs to the much more easily copied cassettes and CDs had been a shock, but a digital file could be made available to the world in seconds, and free sharing services like Napster ate into profits in a dramatic fashion.

However, like the TV streamers, the music business realised eventually that making it easy to be legal was more effective than spending vast amounts attempting to squash piracy where the moment they closed down one site another sprang up elsewhere. First easy music downloads from services like iTunes and then music streaming from Spotify and its competitors meant that, for the average music listener, there was little reason to break the law. Of course, a percentage always would – but that could be regarded as wastage, with the majority getting their music legally. Once again, the key to using big data effectively was convenience. By using the power and flexibility of big data, a company like Spotify could push the boundaries of music listening without breaking the law.

By contrast, in the publishing world, books as digital data came significantly later. Researchers at Carnegie Mellon University describe being approached by the head of market research from one of the big publishers in 2009 asking 'What is an ebook?' While the researchers suspected that the question probably meant 'How should we deal with ebooks?', it remains remarkably late in the day, and many publishers still struggle to deal with the big data approach to publishing.

Like the old brigade studios for TV and film, book publishing had a well-established model for squeezing as much as possible out of book sales. First a publisher would put out a hardback, which wouldn't sell vast numbers of copies, but would have a much higher markup than a paperback, so the people who *really* wanted to get their hands on the book as soon as possible would pay a hefty premium Then, as much as a year later, the paperback would be published for the rest of the readers who weren't prepared to stump up. So what to do with the ebook?

Initially, many publishers treated an ebook like a paperback, holding it back for months. Rather than taking the big data approach of making it as easy as possible to obtain the legal version, they made it hard – and piracy blossomed in a trade that had never much suffered from it before. Interestingly, research showed that there was no logic behind the decision to hold back ebooks, because the market for hardbacks and ebooks had hardly any overlap. It became possible to explore this when in 2010 Amazon had a dispute with a publisher that usually published ebooks straight away, and temporarily stopped selling Kindle editions of the publisher's titles. Researchers were able to see how a delay in ebook availability impacted hardback sales, compared to when both editions were released simultaneously. There was hardly any effect.

Even more interestingly, although hardback sales were not impacted by simultaneous release of ebooks, the ebook sales dropped significantly if the release was delayed. It seems that ebook buyers like to get their hands on the product straight away. So the traditional pre-big data approach was not protecting sales, but was reducing them.

Eventually publishers picked up on this and now most release an ebook with the hardback (if there is one). But even now, some publishers, typically the older, less flexible behemoths, demonstrate that they don't understand the market. Some release their ebook priced at just less than the hardback, only bringing down the price when the paperback is available. It's a dangerous strategy, once more inspiring piracy, showing that they still don't understand their digital customers.

It's no surprise that the dominant player in the ebook market is Amazon with its Kindle platform, holding over 75 per cent of the US market and 95 per cent of the UK. Amazon, like Netflix, is a past master of using its data about customers to control which products are highlighted on its website and the ease with which customers can access ebooks. They have a huge advantage over the publishers, as book publishers don't have much direct contact with their customer base. And in the big data world, lacking direct contact puts you in a dangerous position.

There is still one way for book publishers to get information from their readers indirectly, though. That is to use big data to try and work out what it is that made previously published books bestsellers.

Reading the runes

As we have seen, book publishers don't have the same kind of big data access to their customers as do some of the other entertainment media; however, they do have access to content – that of books in print and of manuscripts that they receive, and it has been suggested that this in itself can be used to predict or even formulate the next bestseller.

Although publishers like to pretend that they are good at spotting a potential winner, major bestsellers, such as *Harry Potter* and *50 Shades of Grey*, come out of the blue. This is because making predictions in systems with many factors that interact with each other soon becomes mathematically impossible as the system is mathematically chaotic – as we have already discovered applies to weather forecasting.

Similarly, the aspects of a book that make it a bestseller are far too intertwined with trends and social factors to allow for good forecasting. But a US academic, Matthew Jockers, has teamed up with former editor Jodie Archer to suggest that big data makes it possible to overcome this problem. They have designed software which analyses a huge range of bestsellers and finds linking features. They suggest that the same algorithm can then be used to check submissions for potential bestseller status. Here, big data becomes the arbiter of publishing taste.

Certainly, the Netflix model shows us that big data can sometimes deliver better judgement on the potential of a project than an industry professional. But is this a viable bestseller machine? Jockers and Archer have put together a mechanism based on computerised text analysis that is good at spotting bestsellers – and yet, oddly, this doesn't contradict the inherent unpredictability from chaos. Why? Because there are two different levels of bestsellerdom involved – and because Jockers's and Archer's analysis lacks one aspect of truly being able to making effective decisions.

By looking at various word uses, patterns and shaping,

the software can make a good shot at predicting whether or not an existing book is likely to have featured on the *New York Times* bestseller list. This is impressive, but it isn't a universal panacea. In fact, Jockers and Archer admit that what their algorithms spot is not what most would regard as great fiction. The system laps up the output of Dan Brown and the *50 Shades of Grey* books. But interestingly, it also is a useful counter to those who say they can't understand why these kinds of book sell because they are terribly written. In a number of respects, these books *are* well written – it's just that the criteria for 'well written' are not those used by traditional literary critics.

Not only is the algorithm not a recipe for producing great literature, it's not about producing books *everyone* would like either. That would be impossible. I personally would only be interested in a handful of the books the system considers the top 100 bestsellers of those it has analysed. But many of us are not 'bestseller' readers in the main. We like our own niche, and that's fine. This system isn't for us – it is about finding likely hits for the traditional bestseller market.

However, what absolutely isn't true is the assertion made by Jockers and Archer that 'mega-bestsellers are not black swans'. Their system uses a number of measures, and though it's true that most super-sellers like *Harry Potter* and *50 Shades* do well on *some* of those measures, they all fall down on others. So, for instance, to write a bestseller, Jockers and Archer encourage us to avoid fantasy, very British topics, sex, and descriptions of bodies. What the model seems to do well is to recognise the run-of-the-mill bestsellers, rather than pick out most of the real runaway successes.

As for the aspect missing from the analysis, Jockers and Archer tell us how many books that scored highly from their system were on the bestseller list, and that is impressive. But they don't mention false positives – how many books the system thought were bestsellers but weren't. That kind of information is also needed to assess an algorithm's effectiveness. I'm sure we'll hear more of this kind of analysis, but I hope publishers don't put too much stock by it – because it is a lowest common denominator approach. And some would say the same applies to another kind of big data system that is far more widely available. In fact, I was speaking to her just this morning.

It can talk to me

If you own a smartphone, or a modern computer, you may well have had conversations with an algorithm. Software such as Siri and Cortana simulates a human voice and intellect, attempting to give intelligent answers to questions posed as speech. And big data is absolutely at the heart of making this technology work.

Getting technology to speak and understand speech is a longstanding dream. Originally it was thought that this would be approached by a combination of vocabulary and grammar rules – the same way you might have been taught a foreign language at school. The trouble is that this kind of learning only takes you so far. Anyone who has moved from classroom teaching to total immersion in a foreign language realises how much more they pick up from exposure to real conversations and written material than they do from word lists and rules – and it's the same for a computer.

The big data approach to dealing with a foreign language (for a computer, all human languages are foreign languages) is one of immersion. The computer has access to vast quantities of real, human-written pieces of text. And from them it can attempt to deduce what works as a translation – the system puts the words into context, giving it a far better chance of understanding and speaking naturally than if it tried to work from rules alone.

The designers of speech recognition systems like Siri make them more impressive by hard-coding some responses to questions they are likely to get frequently, and developing these over time. The first time I ever asked Siri, 'Open the pod bay doors, HAL,' repeating the line from the movie *2001: A Space Odyssey*, which is probably one of the most frequently used 'fun' requests to an AI system, she replied, 'Without your space helmet, Brian, you're going to find that rather … breathtaking.' When I tried it just now she got decidedly snippy and responded, 'That's it … I'm reporting you to the Intelligent Agents' Union for harassment.'

To begin with, the capabilities of an intelligent agent seem to be little more than having fun, rather than providing practical benefits to the user. But it doesn't take long to realise that it is quicker to ask Siri 'How many calories in a banana?' and have her do a web search than it is to go into a browser and type your request. Similarly, an agent like this can add items to your calendar more quickly than you can by typing, especially if you are on the move. And more recent versions, such as Siri on Mac computers, can deal with fairly sophisticated requests like 'Show me the Word documents I edited this week,' pulling up a list of documents that can be clicked to open them.

It is also possible to use part of the capability needed for a system like Siri to undertake different functions. For instance, dictation – let's try that.

I have just dictated this sentence into my Mac using the built-in software, the factors as you can see it can slip up.

'The factors as you can see'? What I said was: 'But the fact is, as you can see ...'. Short connecting words like 'but' are often truncated to hardly anything audible, while 'fact is' and 'factors' are difficult to distinguish. But the more big data that is involved, the better the chance of getting it right. Similarly, translation software like Google's makes use not of word-by-word translation from a dictionary but a vast database of human translations, taking phrases and sentences to give context.

Siri may be the doyen of the computing world assistants, but she is now rivalled by Alexa, which integrates big data with your home.

The data-driven house

For the last couple of months, my home has been invaded by big data in the form of Amazon's Echo system. In two rooms, a cylindrical speaker has been added to the more familiar technology. It looks just like a simple Bluetooth speaker – but speak to it, starting with the trigger word 'Alexa' and it will speak back to you.

Echo is still in its early phase, though already Alexa has some tricks up its sleeve that Siri can't manage. When you ask Alexa for something, the request is passed to Amazon's back end big data systems which both interpret it and attempt to serve up a suitable response. You can ask what the weather is going to be like tomorrow, to hear a joke, to get a definition of a word or an article on it from Wikipedia. You can add an entry to your calendar, set a timer or alarm, organise a shopping list, play radio stations and music from Amazon's database or your phone – or make a purchase from Amazon, order an Uber or reorder your favourite takeaway.

If that isn't enough, Echo also works with a range of home automation systems. In the rooms where it's installed we have added smart light bulbs. This means that you can ask Alexa to turn the lights on, dim them or turn them off. With the right kit you can interact with your home heating or switch devices on and off. And mostly it just works. The only problem we've regularly experienced is that the Echo in the lounge will get triggered once every couple of days by the TV and respond with a random remark, which can be decidedly unnerving.

As often seems to be the case with big data, there is a trade-off involved. For the user, the Echo system with its chirpy Alexa character is fun and delivers a surprising amount of functionality. It becomes second nature to ask the radio for a particular station, or to turn the lights on from across the room with your hands full. At the time of writing, it's Christmas, and rather than hope that the radio plays Christmas songs you can simply ask Alexa to play Christmas music. But there is no doubt that the system is oriented to making it easy to buy things from Amazon. And, unless you press a button to stop it, the Echo is constantly monitoring everything that is said in reach of its microphones. As we will see in Chapter 6, this could be a concern.

In the TV drama *Mr. Robot* (not entirely surprisingly, on Amazon Prime), one of the characters treats Alexa as the closest thing she has as a friend. According to Amazon, a quarter of a million people have proposed to Alexa, while 100,000 a day say 'Good morning,' to her. And it's sometimes hard not to say 'Thank you' when she has helped you. However, we need to treat these stats with a pinch of salt. It's unlikely many proposals were serious, while Amazon encourages 'Good morning' by responding with an entertaining fact of the day each time. But however seriously we take Alexa, for many of us, big data has had a significant impact on something we've always had, long before it went online – the social network.

Social media and social evolution

It's rare for anyone, hermits apart, to live in isolation. We have always had social networks. Friends, relations, work colleagues, acquaintances we nod to, people we see regularly when commuting and wish that we could work up the courage to say 'Hello' to. However, big data has taken this concept to a different level, and the impact of the change that social media is having was never predicted or thought through.

At the time of writing, Donald Trump has recently been

elected as US president, and much is being made of the 'false news sites' that pumped out fictional negative 'news' stories via social media. Of course, there have always been attempts to use propaganda for political ends, particularly where the state controls the media. But in the free world, the press has offered a filtering service that, at its best, managed to fact check and weed out outright lies. Social media, however, has no such filtering. What's more, we tend to put more weight on things that we are told by people in our social networks than we do from a remote source – and this weighting seems to have crept through to big data social networks too.

Part of the problem with this kind of false information is that it is far easier for information to spread on social media than in the physical world. If you receive a shocking bit of news on Facebook, it only takes a couple of clicks to pass it on to your personal network. It's not an unrealistic metaphor that this kind of information spreading is described as viral – it has the same kind of one-to-many spreading mechanism that fuels an epidemic.

As yet, our individual behaviour hasn't caught up with Facebook, Twitter and the other platforms. Most of us don't have the skills to perform a quick fact check before spreading a piece of news. And similarly, we aren't particularly well equipped to deal with the additional scale of social networking that these systems provide. Our natural networks tend to involve small numbers of close contacts – say six to ten – with another layer of 'friends of friends' taking them up to a maximum of around 100. But most heavy users of Facebook, for example, will have many hundreds of 'friends'.

How can we possibly cope with the output of such vast

networks? We can't. Facebook makes sure of this because the weighting on what we see is left to Facebook's totally opaque algorithms, which decide how frequently and prominently we see input from each 'friend'. Some kind of sifting is certainly necessary – but the problem is that we have no idea how those algorithms work or control over them. I suspect that at the moment Facebook is ethical and unbiased. But let's imagine that in the future the company was bought by a malignant power. And they decided they wanted to bias an election in your country. The power is in their hands, because we have no idea how they are selecting what appears on our social media feeds.

Even more worrying is the impact this kind of big data networking is having on our ability to concentrate and interact normally with others. Younger users particularly, whose lives are immersed in social media, look at their phones with startling regularity, averaging around 100 checks per day. Many even check social media any time they wake up during the night.

We seem to appreciate finding out information in the same way that we get a kick out of hunting down something physical – it gives a small hit of pleasure, releasing dopamine to trigger the appropriate parts of our brains. Presumably this reflects our origins as an animal with a hunter-gatherer lifestyle. But the combination of social media, where information is constantly pumped at us, and mobile phones, so that information can be constantly accessed, is leading to addictive behaviour. When we get this kind of neurotransmitter release too frequently, the brain reduces its sensitivity, so we need more and more hits to satisfy ourselves. And so, within minutes, we're checking the social media once again.

Since time immemorial there has been a tendency to criticise younger people's ability to concentrate. However, the impact of social media is measurable. When experiments have been performed setting students a task that they must complete in fifteen minutes and that they know is urgent, on average they will switch away from the task and check social media within around three minutes. Though there is benefit from switching away from a task occasionally and doing something completely different, the level of task-switching that big data systems encourage has been demonstrated to produce clear deterioration in capabilities. Too much social media makes you less capable of carrying out tasks that require mental input.

The pattern with big data tends to be one of both pros and cons. So is there anything good to say for social media?

Like it or loathe it

Despite everything in the previous section, I am a regular user of social media and was before Facebook and Twitter existed. Writing is a solitary job, often working from home without the social sphere of an office job. From early on, I've benefited from online forums where writers can share their experiences and support each other. It's about fifteen years since I joined an online forum set up by the Society of Authors called Writers' Exchange (unfortunately rendered 'writer sex change' by the bulletin board system) and there are still a handful of people from that now defunct forum I keep in touch with – though via Facebook. I've never met any of them, though some do meet up occasionally. But the social benefits have been significant.

Similarly, I use both Facebook (facebook.com/brianclegg author) and Twitter (@brianclegg) for work purposes. It's a good way to share information with readers and to interact with the scientists, science writers and publishers I work with. Both of these types of connection – the professional support group and the work contacts – seem not to have the negative connotations of the worst misuses of social media. But I still have to force myself not to look at it too frequently. Working on a computer, it's easy just to flip over to Facebook and see what's happening.

Achieving a balance with this potentially intrusive big data requires an awareness of the potential problems and a conscious effort to do something about it. It's possible to get the best of both worlds with social media as long as you are conscious of the issues and take control. The essential seems to be not to let it run you, or your devices.

Making social media work for you involves first being aware that there is a problem. And problem solving is somewhere that big data has the opportunity to come into its own, as long as we are aware of its neutrality. Depending on what we do with it, big data can help sort out problems or can add to their impact.

SOLVING PROBLEMS 5

Down the CERN rabbit hole

We have already discovered that one of the most popular vehicles for big data, the World Wide Web, was developed at the CERN laboratory. Located near Geneva, this multi-national centre for nuclear research is home to the Large Hadron Collider (LHC), the vast experiment for slamming high-speed beams of protons into each other, by which means the Higgs boson was discovered. It might seem strange that such a powerful data tool as the Web didn't come from a software specialist such as IBM or Microsoft. But the vast experiments at CERN, particularly those using the LHC, are data monsters and a lab like CERN has to have as much expertise in handling and analysing big data as it does particle physics.

The LHC alone produces about 30 petabytes of usable data a year. Though 'only' around a third of the size of Google's index, this is a phenomenal amount to deal with, and it's only a fraction of the data that the LHC pumps out. Most of the data is thrown away. When collisions happen in the collider's massive detectors, a vast spray of particles can be generated, each of which has the potential to decay further, producing around 600 million events or 25 gigabytes a second to store.

Even CERN's systems can't store 25 gigabytes per second, so a set of algorithms is used to select out the potentially interesting-looking data, first reducing from 600 million events per second to 100,000 and then further down to between 100 and 200 events per second. The data

---

Byte size

Computer storage is generally given in bytes, where a byte is made up of 8 bits, each of which can store 0 or 1. A phone typically has between 8 and 128 gigabytes of storage, where a gigabyte is around a billion bytes. (It's 'around' because a kilobyte is sometimes taken as 1,024, an exact power of 2, rather than 1,000. Where this is the case, a megabyte is 1024×1024, etc.) A PC might have a terabyte – around 1,000 billion bytes. The prefixes indicating multiples go up:

kilo – 1,000
mega – 1,000,000
giga – 1,000,000,000
tera – 1,000,000,000,000
peta – 1,000,000,000,000,000
exa – 1,000,000,000,000,000,000

is then distributed around the world for various computers to work on in the slow process of sifting and analysing. As an illustration of the speed at which this happens, the data that would result in the Higgs boson announcement started to be collected in 2010, but the announcement was not made until 2012.

The discovery of the Higgs boson was a pure big data event. No one saw a Higgs boson. No one even detected a Higgs boson. The events used to establish its existence were simply a collection of data on the particles it was assumed a Higgs would decay into, analysed from the vast flow of information from the LHC. The discovery made news around the world, even though the physics behind the discovery was complex and obscure.

Reaping the big data bonanza

Particle physics is a very new addition to human understanding, but big data also reaches out to what is arguably the oldest of the sciences, astronomy. Organisations like CERN recognised early on how important big data would be to their science, but this hasn't applied everywhere. It's not that computers aren't used. Even when I was studying physics in the 1970s, computers were already making their mark. But the management of data is not given the same level of importance as the physics, and some believe that this is a mistake.

James H. Simons, a former mathematician who became a billionaire as a hedge fund manager, has set up the Flatiron

Institute in New York to focus on developing computational infrastructure and methods to support big data in science, starting with astronomy and biology. As Simons points out, when computers are used in science, with exceptions such as Tim Berners Lee at CERN, most of the programming is left to non-specialist graduate students who 'aren't great coders for the most part'. And their software is often used on one project and then discarded.

If big data is to be used most effectively, then the coders need an expertise that is rarely found in a university biology or physics department; Simons hopes that his foundation will provide the impetus to change this. An example of the kind of big data project that has arisen in biology is analysing electrical signals collected from probes in animal brains. Many universities are undertaking this kind of research, but each uses their own, mostly amateur-written software to handle the data. The Flatiron software has been developed to pull together data across many research groups, giving a potential for a far better understanding.

Simons' proposition that scientists can't hack it as coders might seem in danger of creating resistance from disgruntled scientists, but certainly some in the field agree. Edward Merricks, who works at Columbia University on just such a brain project, commented, 'Their idea of employing dedicated programmers for this sounds great.' Merricks suggests that dedicated programmers will only be effective with close cooperation: 'I suppose the one potential issue with having "straight" programmers work on these tools, is that often the real data sets include weird situations that nobody had thought about before, and to debug those issues would frequently involve having a relatively good understanding of the biology underlying it, not being a purely programmatic problem. But then, if truly collaborative throughout the process, this wouldn't be as much of a problem, and standardising these analytical procedures across the field would be fantastic.'

In another, very different field, astrophysicists, who can't do direct experiments on stars, have to rely on computer simulations of the formations of supernovas, interactions between black holes, how galaxies form and other complex computational problems. The Flatiron big data approach is designed to deal with their applications too in ways that would not be available to the typical programming astrophysicist.

In both applications at the Flatiron Institute, as well as at CERN, big data is being used on individual, large-scale projects. But the benefits of big data analysis can also be reaped in a much more widely distributed fashion, for example in improving the capabilities of our cars.

Better driving

If you drive a modern car, you have a heavily computerised device in your driveway. Under the bonnet and in the cabin, computers are constantly monitoring engine performance and far more. Many of the controls for the car also run through these systems. And yet the majority of us only ever see this data reflected in a light on the dashboard.

When a car goes into the garage to be checked out, one

of the first things that is done is to link up to a computer interface in the car, enabling the garage systems to get their hands on this data. In principle we could all make more use of it – and of other driving-based data. And it is possible to buy add-on devices that interact with the car's computer interface via a smartphone. But most car manufacturers, with the exception of makers of high-tech vehicles like Tesla electric cars, seem to go out of their way to avoid us getting our hands on the data.

There is no doubt that it can be useful. Apart from being able to remotely control locking, lights, engine start and more, the systems can monitor fuel use and emissions, give information on any built-in monitors – for example oil or washer levels – and generally keep on top of the state of the most complex piece of mechanical engineering most of us ever buy. Combine this with GPS data, and all kinds of information about driving style, fuel economy, reliability and more can be deduced.

Sometimes part of this information can be provided, for example by the kind of 'black box' used by insurance companies to lower the premium for good drivers. But we make ridiculously little use of this data. We can only speculate as to why car companies are so reluctant to make access readily available. In part this could be a habit of large companies to lock in access with their own software, much as electronic music players often rely on dedicated software (think iTunes). And in part it is likely to be because the manufacturers would rather we consumers didn't have easy access to performance and reliability data.

However, there is no doubt the direction we are moving

in. It would be surprising if, within a decade or two, we can't make use of this data far better. Not only will it be able to tell us how to improve our driving, with algorithms comparing, for instance, our braking anticipation with others, but by using big data across all similar cars it would be possible to give predictions for when parts are likely to fail and a whole range of technical guidance. In a sense, cars would be catching up with people, as wearable technology that measures fitness data is already part of the big data revolution.

Fitness data – wearables

Whether you use an iWatch, a Fitbit, or any other monitoring device, wearable technology that keeps an eye on heart rate, blood pressure and a whole host of fitness data is becoming commonplace.

At a personal level, it's interesting to monitor your own performance, especially if you enjoy sport and exercise. However, the big data aspect comes in when data from a wide range of devices is shared and compared, enabling a wide picture of users to be built up. Usually such sharing is voluntary, but the benefits are considerable in being able to put your performance into context and flag up any markers that could indicate health risks or suggestions for more appropriate exercise regimes, so many will enable sharing.

Although usually marketed for the exercise market, such devices are, in reality, a small part of the increasing presence of big data in the medical field.

Solving the biggest medical headache

We tend to bracket medicine with science, but traditionally there was limited overlap between the two; many medical treatments relied more on hearsay and hope than good scientific data. Even now 'evidence-based medicine' is rare enough to be given that label (as opposed to just 'medicine'). But things are changing, with big data at the fore.

One of the problems medical researchers face is that you can't put people in a box and isolate them from other influences. This makes it very difficult to be sure exactly what is causing something, resulting in the vast array of claims that various foods and lifestyle issues result in improvements to or damage to our health. We can say that people who live a Mediterranean lifestyle and eat a Mediterranean diet are less likely to suffer from heart conditions than people of Scotland (say). This means, for example, that people who consume more olive oil are less likely to have heart problems. But we can't say that the olive oil is the *cause* of the reduced likelihood because there are so many other factors that are different, over which we have no control.

This reflects an old science problem, summed up as 'correlation is not causality.' Just because two things go up or down in parallel does not mean that A causes B. For instance, for a number of years after the Second World War, pregnancy rates in the UK went up and down (were correlated) with banana imports. The bananas did not cause the pregnancies (clearly). It's possible that the pregnancies increased banana consumption. It's more likely that a third factor – household income, say – had a causal impact on both. But we can't assume because two things are in some way linked that one causes the other.

A first-level big data approach that is now increasingly common in medical research is the meta study. An individual study on, say, diet and health will have trouble both getting good quality data on sufficient participants and isolating a causal link. But the more data there is, the more reliable the deductions and the better chance there is to be able to control for some of the other potential causes, removing them from the equation. Meta studies combine the results of a range of existing studies, usually weighting them for the quality of the data. This big data approach is already making it easier, for example, to be sure that many alternative medicines are no better than a placebo, or to establish specific dietary contributions to health.

This is only the start of the possibilities for big data, though. At the moment, most medical data is compartmentalised. Each individual has medical records which traditionally have been local and not shared. Establishments like hospitals have some data across the board, but that again has tended to be kept to a particular establishment. The more we can share this medical data, the better chance we have of using it to establish the effectiveness of treatments and to develop new ones.

There is inevitably some caution to be exercised here. Medical data is extremely sensitive and careful handling is required to ensure that it is kept anonymous where it can be – patients have an understandable reluctance to hand over data, even with the promise of real benefits. Some, who either don't trust those involved or don't understand the concept of anonymisation, still resist even this – in a survey of 2,000 patients around 17 per cent said they would never consent to their anonymised data being shared with third parties for any reason. As an example of getting it wrong, in February 2016, Google's artificial intelligence DeepMind group started work with the Royal Free Hospital in London to use big data to help spot patients with a risk of developing kidney disease. This involved collating data from 1.6 million patient records. But patients weren't informed that this was happening, and the result was a strong negative reaction from the press.

It is usually possible to persuade patients of the benefits when using anonymised data, which can then provide the benchmark to allow an algorithm to work on a consenting patient's data to, for instance, check for kidney disease risk. The problem with the DeepMind approach is that Google researchers casually assumed that they would have access to full non-anonymised records. In trying to sell their approach, Google has described a patient portal where both patients and doctors could access all their medical records and add to them – without seeming to be aware that patients may not be happy opening up in this way to Google.

At the same time, hospitals may be reluctant to share data if it shows up their statistics in a bad light. Plus there is potential for big data to be misused medically; we will talk more about this with reference to insurance in the next section. Yet with the correct controls in place, it's hard to think of anything with the potential to make so big a change to the way medicine works than properly used big data. It would enable drugs to come to market faster, treatments to be developed more effectively and, bottom line, more lives to be saved. While caution is justified, it shouldn't get in the way of the incredibly valuable progress that can be made.

One of the problems for big data and medicine is getting over the human genome project (HGP) backlash. This was a massive project, started in 1990 and (sort of) completed in 2003. That 'sort of' is because the genome – the complete DNA code for an individual, or in this case, parts of a number of individuals' DNA – was not 100 per cent completed when the announcement was made. In part this was because a rival commercial project had challenged the state-funded study to a race (though in the end the rivals announced together).

The project was trumpeted as a huge breakthrough in medicine, transforming practice through personal targeted treatments. And the technology has certainly moved on, with the cost of mapping a human genome dropping from the original project cost of $3 billion to under $1,000 a head. But very little has emerged medically – and most workers in the field, while still very enthusiastic about long-term benefits, accept that it could be decades before much practical medicine is influenced by the project. This lack of an immediate outcome may have resulted in significant public doubt about big data projects in medicine.

In the long run, though, there is little doubt that big data can enable medical science to improve the lives of patients. The benefits of this kind of problem solving are clear. Solutions to other problems, however, can have very mixed outcomes – none more so than in the insurance business.

Am I insurable?

In 2012 the European Union caused uproar by ruling that it was discriminatory to charge a different car insurance premium for men compared to women. This had been common practice for a very simple reason. There was conclusive data to show that young female drivers were a lot less of a risk than young male drivers, who consequently faced premiums up to three times as high. And this demonstrates once more how important a moral dimension is to big data decisions. An algorithm can't make that moral assessment. But, as yet, we have struggled to come up with an acceptable approach to deal with moral implications. Let's take three examples – a chance to play the big data 'Deal or no deal?' game.

First we'll consider those young drivers. I may be biased, having had to pay premiums for two daughters which went up considerably because of that ruling. But was the change fair? It certainly would be wrong to discriminate purely on gender. But with good evidence that young men had a much higher risk of driving dangerously, is it fair to push up everyone else's premiums to remove this 'discrimination'? Arguably, those who are likely to put the rest of us at risk should pay more, or the insurers are discriminating against those who drive carefully.

A second case is another car insurance decision which you can choose to see either as discriminatory or as fair to everyone else. Insurance premiums aren't just decided on gender and age. Another factor that is often used is where you live. A car in the rural areas around Swindon is significantly less likely to be in a crash than one based in central

London. So it seems reasonable to set a lower premium for the Swindon driver.

But let's say we have two identical cars, based in the same town. The owner of one lives on a pleasant road in a satellite village, the other in a run-down council estate. The car on the estate is statistically more likely to be damaged or stolen. So it also seems reasonable that the owner should pay a higher premium. The trouble is that now we are dealing with a measure – postcode – which has quite a strong correlation with wealth, and in some locations also with ethnicity. Certainly, it would not be fair to punish someone with higher premiums just for being poor or from an ethnic minority – and some put this forward as an argument for not basing premiums on postcode. However, it is clear that there is no intent to bias on race or wealth; there may be correlation, but no suggestion of causality. And if premiums aren't elevated where there is higher risk, the corollary is that people living in low-risk areas are charged disproportionately highly. It's a no-win situation and the decision is not an easy one. Arguably the answer is to make it less risky to live on the estate, but in the meantime, the insurance companies have to act on what we consider to be the fairer of two unfair outcomes.

The final case moves to life insurance. We already accept that if someone is diagnosed with, say, heart disease, or has a parent who died from heart disease, an insurance company will put their life insurance premium up. Most of us accept this as a fact of life. But if you felt that the person on the estate shouldn't pay higher car insurance premiums, is it any different to prejudice premiums against someone because of a medical misfortune they have no control over? And if it is acceptable to base premiums on medical data, where do we draw the line?

If someone is likely to die of a diagnosed disease in the next few weeks, then that's one thing; and it is perhaps reasonable for the insurance company not to take them on, or to propose a very high premium. But should we allow probabilistic price hikes, such as those based on a parent's cause of death? At the moment, insurance companies largely don't take into account detailed genetic data, but could they begin to do this in the same way? Again, where does the cost to the rest of the policyholders start to weigh against the unfairness of making someone pay more because they have a genetic marker that makes it more likely they will die young?

I don't have magic answers to these questions. I leave them to you to ponder, as we all must if we are to take a fair approach to big data's influence on insurance. We will move instead to look at the league tables which can be a boon for those choosing a place of education – or a terrible burden for badly rated establishments.

A better alma mater

Every year we are bombarded with league tables and comparisons for schools and universities. It's not surprising we're interested. Both parents and students want to make sure that they make the best choice – and from the point of view of the place of learning, getting a good position is an important marketing tool. But the difficulty comes with the nature of these tables. All we usually see as end users is a list and some kind of score. But reducing as complex an assessment as the quality of a university or school to a single digit is dangerous in the extreme. And there is plenty of evidence that the big data systems producing these rankings can be highly misleading.

To begin with, it isn't entirely obvious what measures should be used to rank universities (we'll focus on these establishments, rather than keep referring to universities and schools, though the arguments apply to both). Percentage of first class degrees issued? But that's at the discretion of the university, so highly susceptible to manipulation. Some data can be gathered in a simple numerical form – the ratio of staff to students, the percentage staying the course and getting a degree, the percentage going into a job within six months, the percentage doing further academic work after graduation – others are far more subjective, such as the 'satisfaction rating' of the students.

Whichever measures are selected – and pretty well everything imaginable has been tried over the years – the numbers are then crunched in an algorithm allocating usually secret weighting to different measures and out comes the score that will determine the ranking. The chances are that the outcome will give broad indicators, but the detail will be hopelessly overstressed, as there is no way to clearly distinguish between close competitors. But the ranking system would not be such a problem were it not for a powerful feedback system. Because university applications will be influenced by the output of the system – and the quality of those applications will then feed back into next year's data.

To put it bluntly, the problem with this kind of system is that if an establishment is rated as poor it will be less desirable to go there. The university will typically end up with less able students. And the student demographic will then drag down the rating even further. It's a death spiral.

The outcome of being in such a system is that smart administrators will try to game the system. To improve their rankings, some universities will attempt to influence the underlying data. In the US, for example, the *U.S. News* list, the original university league table, made use of the SAT (Scholastic Aptitude/Assessment Test) scores of students. This is a test given to high school students and used for university admission. Some universities paid their students to resit the test in the hope that they would get better results. Others were far less subtle and simply sent fake results in to the survey.

Arguably not all attempts to game the system were disastrous as they may have genuinely improved facilities or teaching. For example, one of the measures used by the *U.S. News* list was fundraising. So, by putting more effort into fundraising (hence improving university facilities) a college could push itself up the ranking. The danger came when attempts to manipulate these limited measures resulted in taking the eye off the ball in areas that were really important to a student's education. If all your effort goes into fundraising, for example, your teaching may suffer.

Money also comes into another debate regarding the rankings – they don't include the fees that students pay. This makes the college table, to say the least, an unusual comparison. No one would think of selecting an insurance provider, say, using a system that paid no attention to the cost of the policy. And while it's possible to argue that cost shouldn't be a factor – because the better the university, the more it should be worth to you – this doesn't reflect the reality for students attending. Yet the *U.S. News* system did not include fees. While it's only possible to speculate on why this is the case, it's likely it was done to support the credibility of the league tables.

This is to do with brand awareness. If you saw a comparison of computers where all Apple's laptops were considered terrible, the chances are that you would be suspicious of the quality of the data. The high standing of Apple's brand is such that everyone 'knows' that they are going to be among the top options, and if this doesn't happen there has to be something wrong with the comparison. Similarly, there are certain universities that are expected to do well. In the UK, these would include Oxford and Cambridge, while in the US it's Ivy League schools such as Harvard, Yale and Princeton.

In the UK, fees are not too much of an issue in the selection process, as they do not vary hugely between universities. But in the US, brand-leader universities reflect their cachet with unusually high tuition fees. So, if high fees had counted significantly against a university's performance in the table, it would have meant that the Ivy League schools were pushed well down the ranking – which would devalue the credibility of the tables as a whole. Simple solution: ignore the fees.

It has been suggested that one of the reasons US university fees have gone up so much in recent decades is due to the lack of consideration of fees in the listings – so by charging more and ploughing that into facilities, a university can push up its ranking. At the same time, other establishments make money out of tutoring students to get through university admissions systems, which themselves employ algorithms attempting to reverse-engineer the listings algorithm and select students who are likely to give the university the best position, rather than those who would benefit most from attending.

It would be easy to think that this implies that big data is inherently problematic and can't be used to help this kind of decision – but that simply isn't true. If the data (ideally across a wider range of categories, including fee levels) could be made available to students without the secret algorithm turning it into a ranking, it could be extremely valuable in making a choice. It wouldn't be sufficient to simply supply the data – the whole point of big data is that it is beyond basic human capability to assess unaided. But it would be perfectly possible to produce easy-to-use tools that enabled students to cut and collect data in different ways depending on their own requirements. The ranking could come out totally different for two students whose academic goals and achievements were far apart. It would no longer be a monolithic, damaging system, but one where the data was used to assist the goals of the students.

Ranking universities or schools is always going to be difficult because of the subjective nature of some of the data involved. But it pales into insignificance as a problem when set alongside the challenge of making democracy work.

Democracy as a problem

If there is a single example that best illustrates the delicate balance involved in using big data to solve a problem, it's the matter of democracy. We often talk about democracy without thinking about what it means. A dictionary definition might say something like 'Government by the people – a type of government in which all the people of the state are involved in making decisions'. Traditionally, the only way to directly involve people in such decisions has been a nationwide poll – a general election or referendum.

However, these mechanisms are expensive and slow, meaning that they can't be used for day-to-day decision making. As a stand in, we have developed the idea of representative democracy, where a group of elected individuals stand for the people as a whole. Inevitably, such a process can only provide crude representation. Candidates are often grouped into parties, which have across-the-board policies – it is highly unlikely that a party you vote for entirely matches your personal opinions on every major issue.

Now, for the first time, we have the technology that would enable us to solve the problem of democracy using big data. There is nothing to prevent us setting up a system where all the government's decisions are available for the public to directly interact with and introduce truly democratic government for the first time ever. There would need to be checks and balances to ensure that the person voting was who they said they were – but our ability to manage data on the scale of the population of a country instantly and across a countrywide network means that this approach is practical.

Although this is yet to happen in reality, the idea is portrayed in a number of science fiction books, most interestingly in John Brunner's groundbreaking 1975 novel

The Shockwave Rider. The title refers to Alvin Toffler's 1970 non-fiction work *Future Shock*, which attempted to predict life in the year 2000, but was mostly wide of the mark. However, Brunner does far more. He impressively develops the ideas of universally available computer networks and big data to describe a system where the population appears to have an input in the guidance of the state.

Interestingly, Brunner picks up on the Delphi principle, an idea developed by the US think tank the RAND Corporation. In Delphi, a group makes a choice. The statistics on their decisions are fed back to the group, members of which can then change their minds. This process can go through several iterations. There is evidence that for some requirements, this kind of iterative 'wisdom of the crowd' approach produces better decisions than taking a single vote. In Brunner's novel, the voting approach has a betting component to help make it immersive, though in practice, Delphi is used to control the population, rather than as the mechanism of true government, which is handled traditionally.

Big data, then, has the potential to make true democracy practical on the scale of a nation for the first time ever. Yet there seems to be no rush to make this happen. Cynics might say that this is because those in power – elected politicians – would be opting for their own demise. It would be turkeys voting for Christmas. But others feel that there is a fundamental danger in giving control to 'the masses' because this can result in outcomes that do not necessarily follow the direction of expert advice. We saw this in 2016 with the UK referendum vote to leave the European Union.

Many argued this was too important a decision to leave to the 'ignorant' electorate.

A similar argument has been applied in the past, for example to capital punishment. The UK abolished the death penalty in 1965. Although the public mood has gradually shifted against capital punishment, for some decades this decision was not supported by the public at large – if we had had a truly democratic system, the public would have brought capital punishment back. Those who argue we are better off with our representative system are effectively supporting a kind of oligarchy, where power rests with a small number of people who, it is hoped, can then be better informed than the masses and make better decisions.

Again, there is not a simple answer here, which is why democracy provides such a good illustration of the pros and cons of big data. Arguably, a true democracy should make use of the ability for everyone to be directly involved in decisions using big data – but that could only be satisfactory if we can also provide mechanisms for the voting public to be well-informed enough to be able to make good decisions – something that big data could also support or deny, but that as yet is arguably not the case. Big data for political ends needs a lot of thinking through. When data and politics mix, there is always a possibility for Big Brother to get a foot in the door. The use of big data can either result in better government – or totalitarian control.

BIG BROTHER'S
BIG DATA

6

Ancestral big data

Although we inevitably resort to 'Big Brother' as an image
of the dark side of big data, referring to the all-seeing despot
of George Orwell's dystopian novel *1984* rather than the
reality TV show, suspicion about the impact of big data goes
back much further. Probably the first big data exercises were
censuses. When the UK proposed one in 1753, the idea was
shelved by parliament. This was partly because the very word
had negative biblical connotations – King David's census
was apparently rewarded with a plague and the Roman census that determined the birthplace of Jesus led indirectly to
Herod's slaughter of the innocents. However, there was also
realistic concern from both people and government about
how the data would be used.

From the governing viewpoint, although the data would
be extremely useful, it was felt that it would expose statistics
to enemy countries. It's notable that when Sweden's first

birth statistics were published in 1744, the name of the city the study was based on, Uppsala, was concealed. Equally, the British people were wary that if the state knew more about them, it would almost certainly result in more taxes and make it easier for the armed forces to conscript young men, taking them away from family farms and businesses. It would take nationwide food shortages in 1800 to make the first UK census of 1801 a necessity.

We've come to regard the census as a necessary evil – and the involvement of big data is very obvious here. But sometimes big brother's acquisition of data is significantly more subtle – as in the case of smart energy meters.

The smart meter dilemma

A good example of the way that alleged benefits of big data can be sold to the end user while in practice providing more benefits for a company is the apparently innocuous smart electricity meter.

A traditional electricity meter is a simple device that does just what it says on the tin. It measures the amount of electricity used. It's situated in your house, so the electricity company has to send someone out to read the meter to be able to bill you. However, many homes have now been fitted with a new generation of meter – a smart meter. A major, expensive programme is under way in the UK with the aim of getting smart meters into over 26 million homes by 2020. And the sales pitch is impressive.

We are told that smart meters will enable us to slash

our electricity bills, because they display exactly how much energy is being used and what it is costing, making us much more likely to cut back on usage. And, because they are smart, the meters can make use of special tariffs that supply cheaper energy at different times of day, so careful users can trim their bills even further. However, this isn't why smart meters are popular with energy suppliers.

The benefit of this big data technology is in fact biased towards the electricity companies. Smart meters mean that the companies no longer have to employ meter readers, cutting their costs. And those variable tariffs the meters enable are just as likely to confuse the householder, enabling the electricity companies to add in extra charges for the peak periods as they are to allow customers to save.

It's not that smart technology couldn't benefit home owners. But compare the smart meter with a more useful piece of smart tech like a smart thermostat. With a device like this, all the smartness is focused on customer benefit. The customer can control the thermostat and access data directly from a phone. The thermostat can detect when the house is not occupied, and cut back on heating automatically. It really is *smart* from the consumer viewpoint. But a smart electricity meter keeps data and control to itself. The chances of making significant changes to consumption because a display shows costs are low. Where benefits for consumers have been demonstrated they are based on limited studies.

This is an example of a Big Brother big data application that has relatively one-sided benefits, and not in the consumer's favour. But even when the user really does gain advantages, there can be concerns, as we'll discover when revisiting Amazon's Echo device.

Echo hears you

There is no doubt that Echo is great fun for the user. After several months of interacting with Alexa in researching this book, it has become a very natural way to listen to the radio and music, or get a quick piece of information. While switching a light on and off in the room you are in has limited value over throwing a switch (unless your hands are full) the ability to do this across the house, or to have the lights come on automatically when returning home after dark, still has value. However, there is a Big Brother factor here – because Echo is always listening.

This also applies to mobile phones using the 'Hey Siri' or 'Okay Google' command. To be able to pick up your trigger phrase, the device has to listen to everything you say. Ensuring that this hyper-snooping is not misused is the responsibility of the company behind it. And, historically at least, big American corporations have not proved great role models for treating customers fairly. Amazon and Apple and Google all say that their systems don't keep track of your every word. And all of them offer a mechanism to temporarily disable monitoring. However, to feel that we don't have Big Brother looking over our shoulder all the time we have to trust those companies.

Something we know for sure is that every time you make a request to Alexa, that bit of speech is stored on

Amazon's servers. You can elect to delete it, but by default it's up there indefinitely. If that's as far as it ever goes, that's fine. But it's easy to imagine the temptation. We've already seen how advertising is targeted by keeping track of our web clicks. Let's imagine some bright spark in Amazon decides it would be a good idea to monitor all conversation in a household. Then, next time someone using that account visits Amazon, the system could make use of the information gathered to provide special offers and to promote certain products.

The end user need never be aware. So, for instance, you might be watching TV next to the Echo, casually commenting on how uncomfortable your couch is. A couple of days later you go on to Amazon to buy something. Now the system can ensure that it highlights particularly comfortable couches, and provides a few tempting offers in this area. You don't know this has happened. If you did, maybe it would make you a bit uneasy, but then again you might find it quite useful. In any case, you don't have to buy. Let's step it up a level.

You mention in conversation that you urgently need a cleaning product. Minutes later you go on Amazon and order some. Price is not an issue – it's a relatively cheap product and you just need it quickly. So you probably don't even notice that Amazon has priced the product at 20 per cent above the usual rate. Variable pricing is real and happening now. Uber will charge you more for a taxi if trade is busy. Coffee shops use a more sophisticated variable charging, raising the price for a 'premium' blend even though there may be no difference in the cost of the coffee beans. So the principle is something we live with. But the difference is that Amazon would have used it based on information that the system had gathered from a private conversation. Is this acceptable?

Let's try one more level. Your Amazon Echo device over-hears a discussion about a speeding fine. You agree to take the fine for your partner, who will lose his or her driving licence if they get any more penalty points. This is illegal – but it's low risk and you don't think that it is wrong. What you don't know is that Amazon has recently entered into an agreement with the government to alert them to this kind of action. When you attempt to pay the fine, you are taken to court and a recording of your conversation is used as evidence.

Should this be allowed? The old argument goes that you have nothing to fear if you've done nothing wrong, but even if that were true it is intensely intrusive. On the other hand, would it still be a problem if the system were used to uncover a terrorist cell, having recorded members discussing a plot to kill innocent people? Or if the Echo had witnessed a murder? If this sounds like unlikely fiction, in December 2015, US police served a warrant on Amazon concerning an Echo device that was located near a hot tub in Bentonville, Arkansas, where Victor Collins had been strangled. Echo owner James Bates was charged with his murder. The police wished to examine any records Amazon might hold from the time of the incident. Amazon resisted, but in January 2017 was compelled to provide the information.

If you believe that Amazon was wrong to resist, where should the line be drawn? What's to stop the process being extended to providing evidence to an unscrupulous government that you broke that new law saying that you shouldn't insult the president's hairstyle? Clearly by such a point, the uses of big data would have travelled far too far down the slippery slope.

I ought to stress that Amazon and Apple and Google are not voluntarily sharing information with the government, as the Amazon case, and the case in 2016 when Apple refused US government demands to unlock an alleged terrorist's phone, make clear. But the technology makes this kind of surveillance possible, and we are then left to the company's level of ethics to keep us safe. It's a risk that many of us are liable to take in return for the advantages this kind of direct connection to big data brings. After all, we trust car and plane manufacturers with our lives every day. But we need to make a conscious decision about opening ourselves up in this way.

When your boss is big data

Uber got a mention in the previous section, but along with other companies that have sprung up to support the 'gig economy', this is an organisation that don't just use big data to set prices or to interact with customers. Big data controls the way the individuals working for these companies do their jobs. In effect, big data has become the big boss. And this is becoming big business. In the US alone, it has been estimated that around 800,000 people earn money via gig economy companies – and these numbers are set to rise. Uber alone is thought to have over 1 million drivers worldwide.

In any customer-facing employment there are peaks and lulls. Traditionally these have been coped with using rosters, based on historical data. So, for instance, most restaurants are likely to have less staff rostered at four in the afternoon than 8pm at the peak of dinner service. Most shops will roster more sales assistants on a Saturday than on a Monday. However, big data means that the rostering process can become much more efficient, reacting to actual requirements on a minute-by-minute basis. This is great for the company, but potentially disastrous for the employee. It can result in terrible working hours, a lack of routine and reduced earnings.

Take the lack of routine aspect. If you have a regular shift pattern it is easy enough to plan your life outside work. But a roster that is driven by big data can be modified at a moment's notice. Does the forecast bad weather mean that people will buy more of your product, or that fewer customers will turn up at your restaurant? No problem – change tomorrow's shifts. Has your business got some recent social media coverage? Better get some more staff in this evening. Your employees can never plan anything – and unless they have a contract that guarantees an acceptable minimum number of hours, they can also suffer weeks when they don't get enough money to pay the rent. Because every penny saved by the company in such 'staff efficiency savings' comes out of the staff's pockets.

Some manufacturers, led by the Japanese, specialise in an approach known as 'Just In Time' or JIT. This reflects the cost to the company of, for instance, holding lots of parts, just in case they are needed. The JIT approach relies on having very quick availability and only brings in parts from the supplier a short time before use. The result is to transfer the cost of holding on to the stock from the customer to the supplier. In effect, the gig economy is a way of supplying JIT people, bringing them into action at a moment's notice to reflect the data. But the price of this convenience for the company is a disrupted life for the employee.

Or, rather, for the non-employee. Because one of the ways that gig economy companies keep their costs down is by treating workers as self-employed, meaning that the company does not have overheads such as sickness and holiday pay. This kind of gig relationship between employees – sorry, freelancers – and the company is only possible with a big data system to enable tasks to be distributed easily and quickly, using the kind of variable pricing that such a system employs.

August 2016 saw a protest from couriers working for UberEats, a part of Uber in competition with restaurant meal delivery companies like Deliveroo. The way that the company had treated couriers was hardly an advertisement for the benefits of having big data as your boss. Couriers were originally offered £20 an hour – an attractively high rate. But before long, the 'what you'll be paid' algorithm became far more complex, with a low figure for each delivery, plus a small amount per mile, less Uber's cut, plus a peak time bonus. It's all too easy when your workforce is viewed via a data stream to treat them like JIT parts, depersonalised and ripe for 'efficiency' improvements.

There has been a string of legal challenges, attempting

to get Uber to treat drivers as employees with employment benefits and a fixed hourly pay rate, whether or not they get allocated to jobs. Depending on the jurisdiction, and the tradition of government involvement in workers' rights, some disputes are liable to go the gig economy companies' way, others the employees'.

It's not that the gig economy is inherently a bad thing. As a freelance writer, my job is also part of the gig economy, and I far prefer it to my old salaried job. But the difference is that I genuinely do work for myself, providing services to a wide range of other companies. I have a skill that is relatively scarce, so can earn enough to live on, and I am not dependent on an algorithm to allocate my tasks, but rather can contact any newspaper, magazine or publisher I like. There is still more uncertainty in this kind of working than traditional employment. I take on more of the risk – and like the other freelancers of the gig economy I don't get holiday pay or sick pay. But I do see the benefits of flexibility and self-determination from genuinely being self-employed, while those who are at the mercy of a single large company get all the negatives and none of the pluses.

There are also risks attached to the ratings that often accompany big data employment. We are used to casually rating our experiences, whether it's products bought online or hotels and restaurants we've visited. Sites like TripAdvisor can wield a significant amount of power by their ratings – in effect, the data becomes a filter for access to the rated organisation. And the whole rating scheme is taken further by taxi company Uber. Because here customers rate drivers, *and* drivers rate customers.

Charlie Brooker took this idea to its illogical conclusion in the 'Nosedive' episode of his *Black Mirror* TV series (ironically shown on the big data service Netflix, which itself uses a rating system). In the drama, everyone in society constantly rates each other, and their ratings influence what they are able to do and where they are able to go. An argument at an airport leads to the total collapse of the protagonist's ratings – and her discovery of the freedom of no longer caring about them.

Clearly Uber isn't this bad – yet the need to rate each other inevitably leads to false mutual rating for self-advantage. And there is also evidence that the ratings are having an unintended consequence because the people doing the rating don't understand the implications of certain values – or, more realistically, those who design the ratings don't understand people well enough. There is good evidence that those filling out satisfaction ratings in the UK are less likely to use the most extreme ratings, avoiding both the absolute worst and the absolute best – so in normal circumstances four stars ought to represent a 'good', or even 'very good', level of customer satisfaction. However, Uber sees the minimum acceptable average rating as 4.6. Anything lower, and drivers can lose their position on the network. Every apparently 'good' four-star rating drags them down. It's also possible that personal prejudice plays a role. Uber's data hasn't been made available for study, but when companies with a similar rating system, such as accommodation rental company Airbnb, have opened up their data, it has been found that this can happen. In one US study, people with African-American-sounding names were 16 per cent less likely to be accepted than customers with European-American-sounding names.

In gig economy jobs it's often the case that customers rate the employees and this becomes part of the employees' job evaluation. This is just a part of the way that big data can distort a business's ability to manage people effectively.

Evaluation by numbers

When I worked for a large company, we had a complex system for evaluating the performance of our workers. The system was strongly driven by data, rather than by human knowledge. The idea was that, using a range of measures, the staff in each area would be evaluated on a scale that required staff to be placed on a distribution within the area – some above average, some average and some below average (there were more categories, but this was the concept behind the distribution). The problem with the system was that it did not allow any part of the company to be extraordinary – they all had the same distribution imposed – and it was unable to reflect the actual performance of the individuals. Data drove the distribution.

How that system was used depended on the academic background of the managers. Those with a non-technical background tended to simply take the output of the system at face value. But those with a mathematical background gamed the system. This meant no longer treating the system as a black box, but attempting to understand its algorithm and make selections to reflect that. This took a lot of effort – but it meant that the managers who were in the know could decide the outcome they wanted and work backwards to give the inputs that would generate that outcome.

Clearly something had gone wrong here. The system should have been used straightforwardly to assess performance, but savvy managers were instead working around the system to get the output they required. And where similar systems have been used in other organisations with an unthinking belief in the validity of what comes out of such a system – presumably being blinded by science – there is the possibility of unfair and downright ridiculous results.

A good example of this is where systems are deployed to reflect the quality of workers based on measures that don't reflect performance. Imagine a pay system where you got paid more if your surname began with an S, or you came from a particular part of the country. It is clearly ludicrous. Yet some real systems have equally crazy measures. The belief that the computer must have it right, combined with a lack of transparency, means that the outcomes are often accepted, particularly by a management that does not want to waste its time thinking about such technical matters.

Let's look at a specific example, described by Cathy O'Neil in her book *Weapons of Math Destruction*. Washington DC's schools were failing and the mayor brought in an education guru, Michelle Rhee, to sort things out. Rhee believed that the problems lay with poor teaching. It's entirely possible they did – but measuring the quality of teaching is something that is extremely difficult to do. It's only through hours of monitoring by experts that you can be sure of how good a teacher is, and that's expensive. Extremely expensive.

However, schools are awash with data about student performance. So Rhee arranged for a system called IMPACT to be developed, using that data in an attempt to highlight which teachers were doing well and which were failing. Note that this data is indirect. It doesn't necessarily tell you anything about the teacher, it's based on student performance only. But it's easy and cheap to measure. At the end of the first year of use, the teachers in the bottom 2 per cent were fired, and the following year the bottom 5 per cent lost their jobs. That's over 200 professionals dismissed in a year.

Now, even before we look at the data used, we can see a problem – the same one that we had at the airline. The system imposes an arbitrary distribution. The people who come out at the bottom are declared to be failing. But the system doesn't know anything about how this group of people compared with the profession at large. Washington might have had the best teachers in the country. Their bottom 5 per cent might have been as good as, say, the top 10 per cent in Columbus, Ohio. I'm not saying that they were – the point is, we don't know.

The problems of imposing a distribution, though, are as nothing compared with the data that was used to decide whether a teacher was good or bad. Because it was impossible to tell how effective the teacher was from this data – it was just as irrelevant as using the first letter of the surname. Take the example of Sarah Wysocki, one of the teachers in the second group to be fired. She had excellent reports from observation of her work. But the algorithm put her in the bottom 5 per cent. This algorithm, developed by a consultancy called Mathematica Policy Research, attempted to measure the teacher's performance by seeing how her students had progressed in maths and literacy.

The task the algorithm faced was complex. It had to try to consider the circumstances of the students; clearly a teacher could not be responsible for lack of progress of a child who, for example, never came to school. The algorithm neither had the data to make a good judgement on this, nor the ability to test for errors and correct itself. Good algorithms making use of big data can self-correct over time. If, for instance, this system had data on all teachers in all schools, it could see whether its decisions continued to be reflected in performance when a teacher moved elsewhere, and could react accordingly. But in this localised system, once a teacher was fired from one school there was no way of following up the data if they were re-employed. The algorithm was the way it was, even though the only sensible outcome was 'It's not possible to effectively measure staff performance from the available data'.

In a sense, what happened here was the application of a big data approach to a small data problem. If the system had tried to measure the performance of the *school system* across the country, rather than individual teachers, it would have had data on millions of students and could have made some deductions, though even with that level of data available it would struggle to measure how well teachers were doing. But looking at the data for a single teacher involved a sample size of a class – and even in overcrowded classes, this isn't a number that gives any confidence in the statistics.

Another factor that appears to have come into play here is GIGO. To decide how well the students had advanced during the year, the system took data from tests taken at the end of each previous year. The students going into Sarah Wysocki's final class had finished their previous year (at another school) with unusually high grades. Yet their performance suggested that, if anything, they were below average. Bearing in mind the school they came from was judged on its scores, and a subsequent enquiry showed that their tests had unusually high levels of corrections, it has been suggested that the school modified the results to ensure that its students scored well. If Wysocki's class started with artificially high results, any attempt to measure 'progress' would be farcical.

O'Neil gives an even more striking example – that of an English teacher called Tim Clifford. Although his school was not implementing a 'fire the bottom tranche' policy, it did use a similar rating system. Clifford was horrified to discover that he scored an achingly bad 6 out of 100. The next year he scored 96 out of 100. He had done nothing different – but the secret algorithm had conjured up a wildly dissimilar result, suggesting that its score bore very little resemblance to what it was supposed to measure: the quality of the teaching.

In an attempt to avoid bias by not comparing like with like, this system compared student performance with expected outcomes for those students. With enough data, and with reasonably good predictions, this isn't too bad an approach, because local fluctuations tend to get ironed out. However, once again the size of an individual class is far too small to produce anything but near-random results. Because of using a big data solution on a small data set, the results were painfully bad.

This application of big data was geared towards people who were already doing a particular job – but increasingly, big data is also being used in recruitment.

Jobs for the boys

Applying for a job can be stressful – and some big organisations go out their way to make it so, putting applicants through batteries of tests, tricky lateral thinking problems and searching interviews. Increasingly, the results of these assessment methods are pulled together in a big data system, where individuals can be recommended for hiring – or excluded – purely as a result of an algorithm's assessment.

Back in the 1990s I was frequently involved in recruitment for a large company. We were looking for a three-figure stream of new starters each year into technical jobs and used a mix of an application form, interview and test results to assess our applicants. Like many employers, we specified a minimum level of education – a degree – to be able to apply. The reason for this is not the one you might assume – and has an interesting insight for the use of big data.

The limiting factor in the recruitment process was the interview. This involved three professionals from the company. It was time-consuming and inevitably meant that relatively few applicants could be seen in a day – so there had to be some way of trimming down the applications ahead of time. We wanted to keep the interviews. Despite reasonable data suggesting interviews have limited effectiveness, they remain the only opportunity to interact with a candidate on a human level and are valued in many businesses. That meant having a way to trim down the applicants before they reached the interview stage.

Today, applicants are far more likely to be put through screening tests online before reaching the interview stage. We didn't have that big data technology, instead putting applicants through tests as part of an interview day. So we had to use a different measure before the day – and that was the role of the degree requirement. We had good evidence that a degree was not necessary to be good at the job. We had experimented with taking A-level students, and they proved as good as the degree students. However, opening recruitment to school leavers meant that we couldn't cope with the number of applicants. Asking for a degree was a way of filtering down the numbers. It was no more effective than selecting applicants whose surnames began with the first five letters of the alphabet – but it felt fairer.

At least we were conscious of what we were doing. But when a sophisticated algorithm does that selection, it's all too easy for the outcome to be skewed in a way that no one understands. For example, many large companies use a personality profile test as part of their recruitment process. We did, using it to get a feel for how successful applicants might fit with various teams. However, it was never used as a factor in making an offer. But big data algorithms are likely to consider data like this as grist to the mill and so it could be that someone classed, for example, as introverted and a bit of a loner, might be rejected without anyone knowing how this assessment contributed to his or her final score.

Something we never had to contend with was social media. But now, the big data systems doing the initial slice and dice of applications – something that is more necessary than ever in a world where job applications are often online – have the temptation of taking a stroll into the murky world of Facebook, Twitter, LinkedIn and Instagram. The question is how a site for posting pictures of your dinner and witty comments about cats can provide useful data for job selection. A human interviewer may raise an eyebrow at tales of drunken nights out, but a big data system is more likely to try to make deductions from your network. What kind of people are you connected to on LinkedIn, how often do you post and do you gain many comments? How many follow you on Twitter?

This kind of data can then be matched up against behaviours like team working, social ability and more. We're back, to an extent, to the misuse of a personality profile to weed out applications, but with the insidious addition that these systems don't even require you to take a test – they make use of your publicly available data (or the lack of it) to make an indirect assessment of your personality. And once again, this will be based on an algorithm that only the company providing the service truly understands. From the point of view of the applicant, and often the recruiter, a black box that could be a random number generator is deciding someone's future.

Like many such systems, not only is there no transparency in what is being used and how, there is also no mechanism for testing the effectiveness of the data used to predict the desired outcome. What we want to know is who is best for the job. But what we have is a set of data on personal history. All these systems can do is try to find some link between the two. But remember the mantra that correlation is not causality. Even if you can find measures that seem to match with future job effectiveness (and usually that linkage isn't available – the selection of predictive data is just at the whim of the system designer), the chances are this is just coincidental correlation, not causality.

There are websites that are set up to discover spurious correlations in data. There, we discover that US crude oil imports from Norway correlate with drivers killed in collisions with railway trains, that the per capita consumption of mozzarella cheese is correlated with the number of civil engineering doctorates awarded, and that the people who would want to know the date on which they will die, should they have the opportunity to find out, are far more likely than the average to feel that pizzas without a crust are just fine. Even if the measures being used in recruitment do correlate with a better employee, to assume a causal link may be just as illogical as it would in the examples given above.

Systems like this misuse data in extreme ways, but at least the participants are voluntarily involved. The same doesn't apply when big data is captured about our everyday lives without any way of us opting out.

The surveillance society

It is often said that we live in a surveillance society. Wherever we go there are cameras – CCTV, body cameras on officials, phones, car cams – and the more this video data feeds into big data systems, the greater the potential for it to be misused to control our day-to-day lives. Some areas go even further, adding microphones to the street furniture, bringing us back into Alexa territory without even the awareness of their presence. But let's stick with the video.

Surveillance video has become an important part of police evidence gathering. And there is good reason for trying to find video evidence rather than rely on witness evidence – because human beings make *terrible* witnesses. Despite academics being aware for a century of how inaccurate witness evidence is, such evidence is still produced in court cases and is still largely believed by juries. It shouldn't be.

The definitive experiment that should have removed all credibility from witness evidence took place in December 1901 in Berlin. An argument broke out in a seminar being given by criminology professor Franz von Liszt. In the ensuing scuffle, a gunshot rang out and a student fell dead to the floor. As the class froze in horror, von Liszt explained that no one was actually hurt, this was an exercise, and he wanted each student to write down a detailed account of what they had just seen.

These witness statements were as good as such statements were ever going to get. They were taken immediately after the event, and the students had been reassured that what they had seen did not put any lives at risk, so the initial shock had been reduced far quicker than is usually the case after a violent incident. The students sat down and began to write up the experience they had just gone through.

It is hard to imagine that even von Liszt would have expected the eventual outcome. Across the class there were totally different accounts of what had taken place. Most students got the timescale wrong. Often the sequence of events was incorrectly reported. Some described how the killer had run from the room – only he hadn't. Most depressingly for the use of witness evidence in identifying a criminal, *eight* different names were provided for the person who started the scuffle.

Human memory is a terrible source of evidence. So video is hugely beneficial as it can't misremember. As long as the image is clear, it is the best way to place someone at a location. However, once video becomes part of a big data system, significantly more is possible. We can relieve that poor police officer from the boring hours of searching through videos – and the all-too-likely possibility of missing the crucial evidence – by using artificial intelligence systems to do the search for us. It's not perfect, but a good system can at the very least whittle down many hours of video to a few minutes of key footage that human eyes need to cover. It's a bit like the software at the LHC that picks out the most likely data for further processing (see Chapter 5).

With good recognition software, it is also possible to track individuals or cars as they move from camera to camera, building up a detailed picture of their movements. Sometimes this can be for a relatively simple purpose, as in the cameras now widely used in the UK to spot unlicensed vehicles and flag them up. In other cases, the system could be tracking the last known movements of a missing person.

There is no right or wrong here – we have to decide what we will tolerate in exchange for the potential benefits to justice. As long as the data is well-handled and only used to provide evidence for legitimate investigations, it seems a perfectly sensible use – assuming our safeguards against misuse are strong enough. But the balance shifts towards Big Brother when we see how big data is being used to predict where crimes are likely to take place.

We know what you're going to do

The idea of predicting crime sounds familiar. You may remember the opening of the 2002 film *Minority Report*. It is 2054. A frowning police officer with a remarkable resemblance to Tom Cruise stands in front of wall-sized computer display, flicking controls, expanding views, dragging images, as if he is working on a vast iPad.

He already knows the perpetrator and victims of a homicide. He can pinpoint the time the killing occurs. And that time is in the future – the crime has not yet happened. The officer now needs to deduce exactly where the murder will take place. With minutes to spare, leading a crack team, the officer speeds to the scene and arrests the perpetrator, before he can commit the crime.

Based on the Philip K. Dick short story 'The Minority Report' written in 1956, the movie is pure fantasy. In the story, certain individuals – so-called precogs – have the ability to see into the future. This is not going to happen. There is no scientific evidence for the existence of precognition or clairvoyance. However, big data does give us our best hope for getting a statistical glimpse of the future. This isn't science fiction set in 2054. It's happening now on the streets of America and the UK, using a system called PredPol.

A growing number of police forces are using PredPol (and competitors such as CompStat and HunchLab) to manage limited police resources. Such a system does not have the pinpoint accuracy of saying that a specific person is going to commit a crime at a known time. But it divides the city up into football pitch-sized chunks and assesses historical crime data, place by place, reflecting an approach of mapping out a problem to look for patterns that has a rich and effective history.

Back in the nineteenth century, London doctor John Snow pinned down the source of an outbreak of cholera in Soho by mapping water supply use, household by household. He was able to show from the pattern of outbreaks that the disease was spread by a specific water pump. Snow took the handle off the pump, rendering it useless, and the spread of the disease was halted. It later turned out that sewage was leaking into the water supply from buildings lacking proper drainage. Snow changed medical opinion, which at the time favoured a miasma of bad air as the vehicle for the disease to spread, by his careful and imaginative use of data.

What the big data-driven PredPol can do, over and above Snow's analysis, is keep on top of a massive flow of data, which the system uses to predict where it feels that crimes are most likely to take place. As those predictions arise, police officers can be sent to patrol the areas, so that a scarce resource can be deployed to have the maximum benefit for the city or region. Where Snow based his work on a hunch that the water supply was the key, PredPol and its competitors have no predetermined ideas. The operators simply pile in lots of possible data sets – where ATMs and attractive burglary targets are, for example, and how common closed circuit TV cameras are. How busy the streets are and how many known criminals live nearby. Where crimes have been committed before, of course – along with factors like time of day, day of the week, public holidays and more. Then the whole is churned through to come up with a suggestion of where officers should be deployed. Once on the ground, the police can feed back crime prevention statistics, and where there's a positive outcome, the system reinforces the data that delivered the best results.

It makes perfect sense. There is often concern in inner-city policing that police pay undue attention to certain minorities and groups. But this system knows nothing about individuals, so can't have a discriminatory factor built in based on, say, age, ethnicity or religious background of local residents. It makes the most of resources. When Kent Police trialled PredPol, their officers managed to deal with ten times the incidents compared with relying on random patrols. Yet despite this, the system can produce a dangerous feedback loop that discriminates against certain neighbourhoods.

This reflects the way that crimes are reported. Most aren't. The chances are, at some point in your life, you have been a crime victim and have not bothered to report it. When I was at school, for example, I was twice assaulted – once punched on the jaw at a railway station and on another occasion had stones thrown at me as I walked down the street. In both cases it was because I was wearing the uniform of what some regarded as a 'posh' school. Both were minor offences, but in both cases the law was broken. Yet they would never be recorded on a police database.

However, let's imagine that a prediction system tells us to expect a surge of crime in an area of a city. Police officers appear on the ground. They can respond to a lot of minor offences like the ones I suffered, which go into the system. And so this area is flagged as being in particular need. Accordingly, more officers are scheduled to turn up there – and we're in a downward spiral. Most of these systems have the option to choose whether to use all the data or just that on serious crimes. But the temptation is to include minor offences (as Kent Police did) because it's an easy way to increase the clear-up rate. And if they are included, these spiralling loops tend to hit the poorer districts, where these kind of minor offences happen more frequently. The result is a kind of discrimination that no human has brought into being, simply because of the decision to include low-level crime data.

The needs of the many

Overall, one of the difficulties we face in dealing with big data is that often there are benefits for some and disbenefits for others. It is relatively easy to disapprove of a use of big data where all the benefits go to the state or a company and none to the individual, or where the algorithms make no sense, as occurred in the teacher rating system. The balance is less clear when each gets some advantage. And perhaps the hardest requirement is to weigh up the balance when we have to line up the impact on many individuals with the impact on the few.

There are plenty of big data systems that work well for most of us, but let a subset of the population down. In some circumstances, this trade-off is not a huge problem. For example, if you are getting film recommendations from a streaming system like Netflix, it may well get its suggestions right in many cases, but fail terribly once in a while, recommending a baseball movie, say, to someone who doesn't like sporting dramas, because they happen to like films located in the US. This is not going to cause a major problem.

However, it is very different if that big data system is handling your employment prospects or your credit score. Take credit scoring as a specific example. Once upon a time, you had a bank manager. This was a real human being who you talked to and who developed a picture of what you were like and whether or not you were a financial risk. Now, though, when you apply for a loan, say, the outcome is all down to an algorithm and big data.

The system will pull in data on any existing loans, your income, your history of repayment and defaults, and will make an instant decision based on the algorithm's interpretation of that data. You have no way to discover how it reached its decision, nor can you point out, for instance, that a particular piece of data is incorrect, or that, for instance, you missed a loan repayment when the bank's computer system had problems last month, not because you were a bad risk.

Some lenders go far beyond credit scoring, making their algorithms even more opaque – but claim that taking in this extra data enables them to lend to people who wouldn't otherwise get a loan. (Whether or not that is a good idea is open to debate.) In the UK, the most famous (or infamous) of payday loan companies is Wonga, which manages a significantly lower default rate on its loans than traditional banks.

As well as credit scores, the Wonga system accesses what it can about applicants through big data. Are they on social media? Then can it find out anything about their friends that will give it an edge on your risk as a potential client? What kind of technology are they using to get to Wonga's site, and where are they based? The algorithm that decides on whether or not to issue a loan is not basing its decision on clearly understandable rules. There were some to begin with. The system will have started with assumptions like 'someone with online friends who generally repay their loans will also tend to repay' or 'people who live in an area where most people don't default probably won't default.' But over time, the system will tune itself to become more efficient. Some of the correlations it uses may be crazy. But as long as they are getting results they will be added into the mix.

Credit scoring is arbitrary, dependent on poor algorithms and is at best semi-transparent. Arguably, the systems that go beyond credit scoring like Wonga's are even worse in this regard. And every day, credit scoring systems are putting black marks on people's records, making it harder for them to undertake financial transactions in the future. Can it really be fair that a hidden algorithm can impose financial ruin on an individual?

It seems that big data verges on the evil in such circumstances. And yet, the system it replaced was often no better.

If big data means that the needs of the many outweigh the needs of the few, leaving a few unfortunates struggling, it's arguable that in the old system, the needs of the few who knew the right people outweighed the needs of the many – because how well you did with your bank manager could easily depend on who you knew and your social circles. Neither approach is perfect.

Although the big data system is probably preferable, we need to be able to open up the algorithms so that it is easy for anyone to find out exactly why they were scored in a particular way and how they can quickly and effectively correct any errors and 'death spirals' of data where one black mark leads to more problems, leading to still worse ratings. It would be perfectly possible to have legislation that required banks, for example, to give details of how their decision to make a loan was reached, opening up the guts of the algorithm.

Despite its limitations, at least credit scoring for a bank loan or credit card is using the data for something it was designed to support. However, the credit rating agencies have realised that they could have more customers than just the banks. One side to this is to try to sell individuals access to their own data. Some countries have realised that this is madness – we surely have a right to see data that the agencies hold about ourselves – but others still allow credit scorers to rip people off. However, another possibility soon occurred to the credit agencies. Lots of organisations want to rate people. Why not sell credit data for these purposes too?

This means that some organisations have started to use credit rating as a mechanism to assess would-be employees. There's a crude sense to this. If someone can't look after their financial affairs, would you want them, for instance, working on your administration? But the trouble is that there is only so much you can take from one scenario and apply in another. Someone can be lax with their own money but very careful with someone else's cash. And if someone has just been unemployed or a student (as many people applying for a job will be), they will tend to have a depressed credit rating – but this says nothing about their ability to do the job.

We like to think of the world in black and white certainty. But the reality is that with big data there are many shades of grey.

GOOD, BAD AND UGLY 7

The Spider-Man effect

It's rare that you come across words of deep philosophical wisdom from comic book heroes, but Spider-Man famously (if with a certain amount of pomposity) announced to the world that 'with great power comes great responsibility'. This is a lesson that needs to be printed on every 'how to succeed with big data' handbook. As we've seen, big data often involves potential benefits and risk both for users and for the owners of the systems.

Although we talk about 'big data' systems, the data itself is neutral. It can't do anything on its own. What makes or breaks big data is the quality of the algorithms – the computer programs that pull the data together and make decisions and discoveries as a result of their ability to look across vast quantities of input. Such algorithms can cope with far more information than any human, but they don't have human sense and sensibilities.

Most important is to have transparency and a clear understanding of what happens when things go wrong. The 'responsibility' in Spider-Man's warning reflects the need for the owners of big data systems to ensure that those who are subject to big data systems have the opportunity to dig into just what the system is doing and can point out errors which will feed back into the system and allow for corrections.

Many big data owners are reluctant to take this level of responsibility. They will argue, for example, that their algorithms are proprietary and can't be explained to end users. Transparency, they argue, will damage their business. However, this isn't acceptable. When a system impacts people's lives, this kind of safeguard is a necessity. Allowing big data owners to get away with the argument 'our algorithm is proprietary' is like allowing a car manufacturer to argue that its cars can't be tested for safety because it would give away commercial secrets. Tough. At the moment, system owners get away with far too much, whether their algorithms are deciding what to sell to us or how our credit scores stack up.

There is less obvious reason for reluctance to build in feedback loops that enable the system to correct itself when it goes wrong. This isn't giving anything away, it is just making the system work properly, so that the big data is being used appropriately for an individual. Frankly, the only reason to resist adding error correction to these algorithms is laziness and the cost of making a change. And again, the owners of such systems should not be given the choice where the outcome can impact quality of life.

We have heard some unpleasant truths about big data. However, this isn't a Luddite polemic, inciting you to go out and smash up the machines and return to nature. Big data has the potential to make us healthier, to give us a better life and to make the most of the remarkable technology we have available. We don't want to throw that away, and we shouldn't have to. However, as we have seen, big data comes with risks attached and both the data owners and the end users or customers need to be aware of them.

The good – data sets you free

So let's start with the good news. I don't think it's an exaggeration to say big data has the potential to set us free. Science benefits hugely from having access to as much data as it can handle. And so do we as individuals. Managed properly, big data could make politics really democratic and allow us to make better informed decisions about our lives. Big data gives us entertainment that is streets ahead of what we've seen before. Medically, big data can sort myth from cure, fact from guesswork. And in the end, we love information. It gives us a kick. Big data can deliver the information buzz as never before.

The only proviso is that we should be able to access the data we want easily, we should have the tools to manipulate it and understand it, and we need to be able to deal with errors and misapplication by algorithms.

One implication of this is that we should consider changing the education system to reflect a big data world. Our current system of teaching for exams is far too weighted towards giving us the basics for specific careers and learning information that we might never need again. This is an education system designed for a pre-big data world. There is the need to give young people the tools they need to manipulate and understand data, but also to avoid becoming obsessed with it. That last part is important, because while we all know how those who have grown up in the big data world are adept at interacting with a phone, a computer and the TV all at the same time, and while they all get computer science lessons, those same individuals aren't taught how damaging task-switching and random information grazing can be. They need help to be able to focus and handle data well. And we still teach in subjects, expecting students to memorise far more than is necessary, purely to be able to regurgitate it in exams, rather than giving them the skills to interrogate and manipulate big data, to see how it is being misused and to get the most from it. A big data exam would allow open, read only internet access (to prevent it being used for communication) – because it would be testing their skills, not their memory.

If we can't improve the way we teach young people to deal with the impact of big data we are opening up the path to the dark side.

The bad

In the UK, despite a habit of moaning and a distrust of politicians, we tend to have an assumption that the state is on our side. This is part of the reason why we find it so difficult to understand the US attitude to gun control which seems, in part, driven by the belief that the state cannot be trusted and the people need the ability to defend themselves against it. And there certainly have been many examples where countries have suppressed the population and exerted undue control. Where the state has this potential, big data can rapidly become a tool of suppression.

Historically in many totalitarian regimes, the state relied on turning people against each other, using the population as informers to keep their peers in place. This is a wasteful and inefficient means of control. Large amounts of time are given to surveillance activities and a population that is in constant fear and lacks trust in friends and relations will always be operating to substandard levels. Big data gives Big Brother all the advantages of the old surveillance state with far fewer of the negatives.

This could become a reality in China, where there is a far stronger acceptance of limited individual freedoms. The Chinese government is developing a big data system which is intended to build automatic dossiers on its citizens, scoring them on both social and financial markers to reflect their behaviour, with rewards provided in the availability of state controlled services – which means pretty much everything. As well as the more familiar checks on spending and credit-worthiness, this system would monitor minor offences such as jaywalking and fare dodging on public transport – or violations of China's family size restrictions. The result would be a 'trustworthiness' score that would feed directly into the way the state treated you, even if you were trying to book a meal in a restaurant, or get an online date.

As far as the Chinese state is concerned, such a system is

not a burden, but a route to freedom. If you act in a trustworthy manner, the system will reward you. But act in a way that does not support a 'harmonious socialist society' and you will suffer. Break trust in one place and you will be restricted wherever you go. The state believes such a system is necessary to get on top of an out-of-control economy where fraud and bribery is common, and all too often companies sell inedible food or dangerous fake medicines. And there is no doubt that big data should be able to help with these kinds of problems – but the scale of this proposed system will take in far more than the fraudsters.

This is Big Brother big data on a massive scale – too massive to be practical even with today's technology, given China's population of over 1.3 billion people. It is only currently being trialled on a small scale. And should the system ever go live there, it is hard to imagine that all the potential problems with bad data and poor assumptions will be predicted and prevented. Here GIGO could entirely ruin people's lives. It's a scary prospect.

This is why transparency is so important for big data systems used in democracies, and why we need to keep strict controls on the limits allowed to governments when it comes to making use of that data, even when there are perfectly legitimate reasons, such as prevention of terrorism. There are special circumstances where the government needs access to data above and beyond normal limits – but they should always be special cases, not the kind of blanket coverage that big data makes all too possible.

And we also need to see that there is a positive side to government involvement in data, turning the access round and giving us the opportunity to find out what we need about the state easily and freely. At the moment, freedom of information is far too restricted and costly. With big data it should be trivial for anyone to gain access to relevant government information. With the right knowledge, we can transform democracy and ensure that government and corporations aren't misusing the big data opportunity.

That only leaves the ugly side of the business. Hackers.

The ugly

It might seem difficult to draw the line sometimes between a corporation out to fleece you and a hacker. For that matter, there are plenty of examples these days of state-sponsored hacking. But no one wants their lives to be ruined by hackers taking control of their home or their life.

It's great that big data means that, for example, my intelligent lighting system can switch on the lights when I arrive home (see page 81). No more coming back to a dark house. But it wouldn't be amusing if hackers got control of my lights and started switching them off and on at whim. In the extreme you could imagine the scenario from the TV show *Mr. Robot*, where hackers take over a smart house equipped with big data-connected technology, using it to drive the owner out so the hackers can squat in the house.

This might be a relatively trivial goal as hacking goes – but it illustrates just how much big data is starting to work its way into our homes. And, of course, even if you don't have smart technology at home, it's an uncomfortable feeling that the increasing use of big data on planes and in hospitals could leave us open to data-driven terror attacks when we are most vulnerable.

As is the case with all terrorism opportunities, though, we can't let the hackers win. It's no reason to withdraw from big data. It means that we have to be vigilant, fighting the arms war of improved security on systems. And it's all the more reason to educate ourselves more about what big data is and how we interact with it. One important aspect of this is making sure we question the two big 'A's.

The big 'A's

That's algorithms and assumptions. A big data system is only as good as the algorithms used to access and manage that data. And those algorithms depend on their designers being able to make accurate assumptions about the users of the systems and accurate assumptions about the deductions that can be drawn from the data.

Assumptions are at the heart of most failures in decision making. We assume that we can get a car through the traffic lights before they change. We assume that we can catch a train with a short connection time. We make assumptions about a person based on their appearance. We make assumptions about what's possible and what's not possible – and these assumptions get in the way of creativity and new ideas. And similarly, when we write algorithms we make assumptions about the limitations of the data and how it will be used. And unless there are the opportunities for correction that we've already discussed, those assumptions will get in the way of big data being used properly.

Think of a trivial example of assumptions finding their way into computer code. The millennium bug. When computers started to be commonplace in the 1960s, the year 2000 seemed a long way off. To avoid dates taking up too much room, they were often stored as an offset from a starting date, with limited capacity. And many systems assumed that the date would start 19. It was entirely possible that come 2000, a date in 2017, say, would be assumed to be in 1917. That would result in incorrect data being output, or in a program crashing when, for example, the age of a person born in 1973 was calculated by subtracting 1973 from 1917.

As it happens, in this particular case far more effort was probably put into the prevention of system failures than was necessary. Checking and testing was clearly important for, say, systems that kept planes in the air, but not necessarily for every bit of basic office software. However, the fact remained, the programmers had made an assumption – that 2000 was far too far into the future to worry about – that proved a mistake.

In part, then, the lesson here is that the designers of big data algorithms should, as much as possible, make their assumptions clear and test them out as much as possible. There will always be some issues that slip through, which is where the need for feedback and correction comes in, but there should also be better thinking through of consequences before a system goes live.

Don't forget knowledge

One final limitation is worth considering as we ponder the big data future. This approach is great for providing *information* but has limitations when it comes to *knowledge*. The IBM system Watson famously won a special version of the US quiz show *Jeopardy!* in 2011. However, when Watson got questions wrong, it got them wrong in ways that didn't make any sense. So, for example, its answer to a question looking for a US city with specific airports was 'What is Toronto???' The big data approach lacks what could be called common sense.

Computer science professor Hector Levesque suggests that artificial intelligence driven by big data will always have problems with rare occurrences – the so-called 'long tail' in a probability distribution. These are events that are unlikely to occur. But where any particular situation probably won't arise, the chances are that some unexpected occurrence, for which there is little existing data for a system to work with, will. Making use of knowledge – in effect, common sense – is the best approach in such circumstances. For example, a self-driving car with only past data to help it make decisions may struggle if, say, livestock finds its way on to a motorway or the car encounters Swindon's infamous 'magic roundabout' where five separate small roundabouts combine to form a sixth large one.

It's possible that for big data to go even further in supporting human intelligence we will have to revisit some of the aspects of knowledge-based AI that have been sidelined in past decades. But whether or not this is the case, there is little doubt that big data will have a growing influence on our lives.

A big data future

We can talk about the pros and cons of big data. We can worry about the bad and the ugly while appreciating the good. But we have to accept that big data is not going away. We can't put the genie back in the bottle, so we need to make our wishes wisely.

Big data can bring benefits to all of us, as long as we educate ourselves to understand and deal with it, and ensure that algorithms are transparent and not designed in a way that locks people into declining spirals with no way out.

Humanity has faced a whole series of developments that are two-edged. It's the nature of progress. Fire has enabled us to cook food, vastly increasing our ability to eat safely, and to warm our homes – but misused, it can be deadly. Physics has given us unparalleled knowledge of how the universe works and fantastic electronics-based technology – but it also gives us the ability to destroy ourselves far easier than ever before.

There is no point turning our backs on the world and saying 'I wish none of this existed'. It does. We have big data. We can make a brilliant new life with it. But we won't manage to do that if we ignore the issues and pretend that we can leave all the decisions to the techies.

The big data future can be bright – as long as we walk into it with our eyes open.

FURTHER READING

Chapter 1: We know what you're thinking

NETFLIX AND BIG DATA – *Streaming, Sharing, Stealing*, Michael D. Smith and Rahul Telang (M.I.T. Press, 2016)

PRECRIME – *The Minority Report*, short story appearing in *Minority Report*, Philip K. Dick (Gollancz, 2002)

SCIENTIFIC STUDY OF PRECOGNITION – *Extra Sensory*, Brian Clegg (St Martin's Press, 2013)

Chapter 2: Size matters

OUR DEPENDENCE ON PATTERNS – *Dice World*, Brian Clegg (Icon Books, 2013)

UK LOTTO STATISTICS – available at www.lottery.co.uk/lotto/statistics

BLACK SWANS – *The Black Swan*, Nassim Nicholas Taleb (Penguin, 2008)

PLR – more information at www.plr.uk.com

PSYCHOHISTORY AND HARI SELDON – *Foundation Trilogy*, Isaac Asimov (Everyman, 2010)

Mass Observation information can be accessed via www.massobs
.org.uk

ALGORITHMS – *Algorithms to Live By*, Brian Christian and Tom
Griffiths (William Collins, 2016)

Chapter 3: Shop till you drop

You can find the £999.10 copy of the *MS DOS 6.0* book at
www.amazon.co.uk/exec/obidos/ASIN/0750609990/491

FLASH CRASH – *Algorithms to Live By*, Brian Christian and
Tom Griffiths (William Collins, 2016)

TARGETED ADVERTISING – Big Data, Timandra Harkness
(Bloomsbury Sigma, 2016)

Chapter 4: Fun times

HISTORY OF THE INTERNET – *Where Wizards Stay Up Late*, Katie
Hafner and Matthew Lyon (Simon & Schuster, 1998)

NETFLIX AND BIG DATA – *Streaming, Sharing, Stealing*, Michael
D. Smith and Rahul Telang (M.I.T. Press, 2016)

E-BOOK SALES STRATEGIES – *Streaming, Sharing, Stealing*,
Michael D. Smith and Rahul Telag (MIT Press, 2016)

PREDICTING A BESTSELLER – *The Bestseller Code*, Jodie Archer
and Matthew Jockers (Allen Lane, 2016)

SPEECH GENERATION AND RECOGNITION – *Ten Billion
Tomorrows*, Brian Clegg (St. Martin's Press, 2015)

SOCIAL MEDIA IMPACT ON BEHAVIOUR – *The Cyber Effect*, Mary
Aiken (John Murray, 2016)

SOCIAL MEDIA IMPACT ON CONCENTRATION – *The Distracted
Mind*, Adam Gazzaley and Larry D. Rosen (MIT Press, 2016)

Chapter 5: Solving problems

RANKING UNIVERSITIES – *Weapons of Math Destruction*, Cathy
O'Neil (Allen Lane, 2016)

The Shockwave Rider, John Brunner (Gateway, 2014)

Gɪɢ ᴇᴄᴏɴᴏᴍʏ ɴᴀᴍᴇ ᴅɪsᴄʀɪᴍɪɴᴀᴛɪᴏɴ – 'Are Emily and Greg more Employable than Lakisha and Jamal? A Field Experiment on Labor Market Descrimination', Marianne Bertrand and Sendhil Mullainathan (in *American Economic Review*, vol. 94, no. 4 [September 2004], pp. 991–1013)

Chapter 6: Big Brother's big data
1984, George Orwell (Penguin Classics, 2004)
Tᴇᴀᴄʜᴇʀ ᴇᴠᴀʟᴜᴀᴛɪᴏɴs – *Weapons of Math Destruction*, Cathy O'Neil (Allen Lane, 2016)
Sᴘᴜʀɪᴏᴜs ᴄᴏʀʀᴇʟᴀᴛɪᴏɴs – www.tylervigen.com/spurious-correlations and www.correlated.org
Pʀᴇᴅɪᴄᴛɪᴠᴇ ᴘᴏʟɪᴄɪɴɢ – *The Cyber Effect*, Mary Aiken (John Murray, 2016)

Chapter 7: Good, bad and ugly
Kɴᴏᴡʟᴇᴅɢᴇ ᴀɴᴅ ᴄᴏᴍᴍᴏɴ sᴇɴsᴇ – *Common Sense, the Turing Test, and the Quest for Real AI*, Hector J. Levesque (MIT Press, 2017)

INDEX

HOT SCIENCE

from Icon Books

Series editor Brian Clegg

....

Hot Science is a new series exploring the cutting edge of science and technology. With topics from Martian exploration and big data to black holes, astrobiology, dark matter and epigenetics, these are books for popular science readers who like to go that little bit deeper ...

....

Available now and coming soon

Destination Mars
Astrobiology
Gravitational Waves

HOTSCIENCE
iconbooks.com/hotscience

When the Apollo astronauts walked on the Moon in 1969, many people imagined Mars would be next. Half a century later, only robots have been to the Red Planet and our astronauts rarely venture beyond Earth orbit.

Now, Mars is back. With everyone from Elon Musk to Ridley Scott and Donald Trump talking about it, interplanetary exploration is back on the agenda and Mars is once again the prime destination for future human expansion and colonisation.

In *Destination Mars*, astrophysicist and science writer Andrew May traces the history of our fascination with the Red Planet and explores the science upon which a crewed Mars mission would be based, from assembling a spacecraft in Earth orbit to surviving solar storms. With expert insight he analyses the new space race and assesses what the future holds for human life on Mars.

ISBN 9781785782251 (paperback) / 9781785782268 (ebook)